SOUTH-EAST ASIA
WAN
ON
PHILIPPINES
VISAYAS
MINDANAO
STRAITS
OF
MAKASSAR
SULAWESI
MOLUCCAS
E S I A
IRIAN JAYA
KAI
ISLANDS
TIMOR

IMAGES OF ASIA

Series Adviser: SYLVIA FRASER-LU

Bamboo and Rattan

Titles in the series

At the Chinese Table
T. C. LAI

Balinese Paintings (2nd ed.)
A. A. M. DJELANTIK

Bamboo and Rattan: Traditional Uses and Beliefs
JACQUELINE M. PIPER

The Birds of Java and Bali
DEREK HOLMES and STEPHEN NASH

The Birds of Sumatra and Kalimantan
DEREK HOLMES and STEPHEN NASH

Borobudur (2nd ed.)
JACQUES DUMARÇAY

The Chinese House: Craft, Symbol, and the Folk Tradition
RONALD G. KNAPP

Chinese Jade
JOAN HARTMAN-GOLDSMITH

Early Maps of South-East Asia (2nd ed.)
R. T. FELL

Folk Pottery in South-East Asia
DAWN F. ROONEY

Fruits of South-East Asia: Facts and Folklore
JACQUELINE M. PIPER

A Garden of Eden: Plant Life in South-East Asia
WENDY VEEVERS-CARTER

The House in South-East Asia
JACQUES DUMARÇAY

Images of the Buddha in Thailand
DOROTHY H. FICKLE

Indonesian Batik: Processes, Patterns and Places
SYLVIA FRASER-LU

Japanese Cinema: An Introduction
DONALD RICHIE

The Kris: Mystic Weapon of the Malay World (2nd ed.)
EDWARD FREY

Life in the Javanese Kraton
AART VAN BEEK

Macau
CESAR GUILLEN-NUÑEZ

Mammals of South-East Asia (2nd ed.)
EARL OF CRANBROOK

Mandarin Squares: Mandarins and their Insignia
VALERY M. GARRETT

The Ming Tombs
ANN PALUDAN

Musical Instruments of South-East Asia
ERIC TAYLOR

Old Bangkok
MICHAEL SMITHIES

Old Manila
RAMÓN MA. ZARAGOZA

Old Penang
SARNIA HAYES HOYT

Old Singapore
MAYA JAYAPAL

Sarawak Crafts: Methods, Materials, and Motifs
HEIDI MUNAN

Silverware of South-East Asia
SYLVIA FRASER-LU

Songbirds in Singapore: The Growth of a Pastime
LESLEY LAYTON

Traditional Chinese Clothing in Hong Kong and South China 1840–1980
VALERY M. GARRETT

Bamboo and Rattan

Traditional Uses and Beliefs

JACQUELINE M. PIPER

SINGAPORE
OXFORD UNIVERSITY PRESS
OXFORD NEW YORK
1992

Oxford University Press

Oxford New York Toronto
Delhi Bombay Calcutta Madras Karachi
Petaling Jaya Singapore Hong Kong Tokyo
Nairobi Dar es Salaam Cape Town
Melbourne Auckland
and associated companies in
Berlin Ibadan

Oxford is a trade mark of Oxford University Press

Published in the United States by
Oxford University Press, Inc., New York

ISBN 0 19 588998 3

British Library Cataloguing in Publication Data

Piper, Jacqueline M.
bamboo and rattan: traditional uses and beliefs.—(Images of Asia)
I. Title II. Series
584.93

ISBN 0–19–588998–3

Library of Congress Cataloging-in-Publication Data

Piper, Jacqueline M.
Bamboo and rattan: traditional uses and beliefs/Jacqueline M. Piper.
p. cm.—(Images of Asia)
Includes bibliographical references and index.
ISBN 0–19–588998–3:
1. Bamboo—Social aspects—Asia, Southeastern. 2. Rattan—Social aspects—Asia, Southeastern. 3. Bamboo—Asia, Southeastern—Folklore. 4. Rattan—Asia, Southeastern—Folklore. 5. Ethnobotany—Asia, Southeastern. 6. Material culture—Asia, Southeastern. 7. Asia, Southeastern—Social life and customs.
I. Title II. Series.
GT5899.B36P57 1992
394'.36—dc20
91–31343
CIP

Printed in Singapore by Kyodo Printing Co. (S) Pte. Ltd.
Published by Oxford University Press Pte. Ltd.,
Unit 221, Ubi Avenue 4, Singapore 1440

For Isobel and Roland

Preface

SOUTH-EAST ASIA is home to a large proportion of the world's tropical plant species, plants which produce food, fibres, and medicines. Any listing of useful South-East Asian plant species, however, would have to start with two canes, bamboo and rattan, which fulfil so many purposes that to list them all might be tedious. The two canes are found across the region, either growing or in use. The visitor and resident in South-East Asia come into contact with these materials many times a day: walking or sitting on them, eating them or eating with them, and working with them.

This book tries to bring some order to this multiplicity of uses, to describe them and to explain why bamboo and rattan are so widely the preferred materials of craftsmen, artists, fishermen, and farmers.

After a brief introduction to the canes in Chapter 1, comparing their essential characteristics and structure, the following four chapters cover the range of practical uses. Chapter 2 describes how canes are used in building and furnishing homes and shelters and for everyday life. Agricultural and ornamental uses are also indicated. Chapter 3 looks at instances where the plants are consumed either for food, medicine, or used for magical purposes.

The great range of arts and handicrafts in which canes are involved because of their resilience, flexibility, and strength is introduced in Chapter 4. The forms of the two canes have been interpreted by artists across the region: in designs for textiles and pottery and in paintings. The world outside the home also requires canes: for transport, for fishing and hunting, and for fighting battles. These are described briefly in Chapter 5.

Chapter 6 turns away from the practicalities towards culture, myth, and entertainment. Folk beliefs, music-making, toys, and rites of passage all lean heavily on bamboo or rattan canes. Bamboo has a role to play in the birth of children, in courtship, and in death—and rattan is never far away.

The last chapter provides some information on botanical and silvicultural aspects of producing bamboo and rattan and introduces their natural history in more detail for those who wish to understand more of the structure and origins of these materials.

Wolfson College, Oxford JACQUELINE M. PIPER
June 1991

Acknowledgements

I wish to thank the following people and institutions for their kindness in permitting me to reproduce the following illustrations:

Bodleian Library, Oxford, for Plate 22
Department of Plant Sciences, University of Oxford, for Plates 1, 11, 27, and 28, and for the help of library staff
John Dransfield for Colour Plates 8, 12, 23, and 24 and Plates 4 and 30
Sylvia Fraser-Lu for Colour Plate 18
David Goddard of the International Sailing Craft Association (ISCA), Exeter, for Plates 20 and 21
Betty Gosling for Plate 23
E. Heins of the Etnomusicologisch Centrum 'Jaap Kunst', Amsterdam, for Colour Plate 22 and Plate 26
Greg Irvine for Colour Plate 1
Sumet Jumsai for Plate 7
David Kirby for Colour Plate 13
Walter and Mona Lutz for Colour Plates 14, 15, 16, and 17
Hedda Morrison for Plates 5, 6, 9, 10, 15, 17, 18, 19, and 29
Oxfam, Oxford, for Colour Plate 20
Penerbitan Yayasan Kanisius, Yogyakarta, for Colour Plate 21
Brian Pope for Colour Plates 7, 9, and 10, and Plates 8 and 13
Grace I. Selvanayagam for Colour Plate 19
Tatiana, Singapore, for Colour Plate 11
Wendy Veevers-Carter for Colour Plate 2
Roxana Waterson for Colour Plate 6

I should also like to record my thanks to Ben Piper for his help and support.

Contents

I
Tree Grass and Scrambling Palm

IMAGINE what would be the ideal material for making all manner of things: it would have to be strong yet light, durable but biodegradable, smooth to the touch, easy to work, cheap, and always near at hand. Imagine also that this ideal material can be transformed, in the hands of creative people, into objects of beauty, even when used for mundane tasks. In South-East Asia two group of plants grow which possess just these properties. One is hollow—bamboo—and one is solid—rattan. Very many practical human needs can be answered either by a bamboo or by a rattan and often they are used together, as in house-building and boat-building. The great number of species in both plant groups provides a diversity of sizes which means this ideal raw material is available in many different diameters and lengths.

Although bamboos are classified with the grasses, the characteristic all bamboos share is that they lay down wood in their stems which gives strength and makes great size attainable. Some grow as tall as 30–40 m and reach a diameter of 25 cm. Whatever their size, most are well-suited to practical tasks. South-East Asia is the home of about half of the world's 700 or so species of bamboo. Some bamboo species are found in the Americas and Africa but there are no native bamboos in Western Asia and Europe. In South-East Asia the bamboos fulfil a range of uses that stretches, in a man's life, from cradle to coffin, providing food, shelter, tools, and entertainment on the way.

In order to understand the wealth of opportunities offered by the bamboos and the rattans, it is helpful to know a little of their physical structure.

Like almost all grasses, the bamboo's stem (or 'culm') is cylindrical and hollow. The stem is divided into segments by cross-walls which separate the stem into self-contained sections called joints or internodes. On the outside, narrow ridges (known as nodes) encircle the stem at regular intervals, marking the location of the cross-walls (Figure 1). Instead of growing like a tree, at the top, bamboos grow by extending each joint upwards from its base. As each of the 20–40 individual joints of a stem may be growing at any one time, very fast growth is achieved: 75 cm per day has been recorded though 30 cm in 24 hours is more common. Bamboos grow nearly as fast at night as in the daytime.

Early Western travellers in the East found bamboo paradoxical. In his *A New Historical Relation of the Kingdom of Siam*, Simon de la Loubère, envoy of the King of France in 1687, gave this description: 'One of their most eminent Trees is a kind of Reed, called in Indian, Mambou, in Portuguese, Bambou and in Siamese, Mai pai. This Tree resembles the Poplar, it is strait and tall and the Leaves thereof few, pale and longish.'

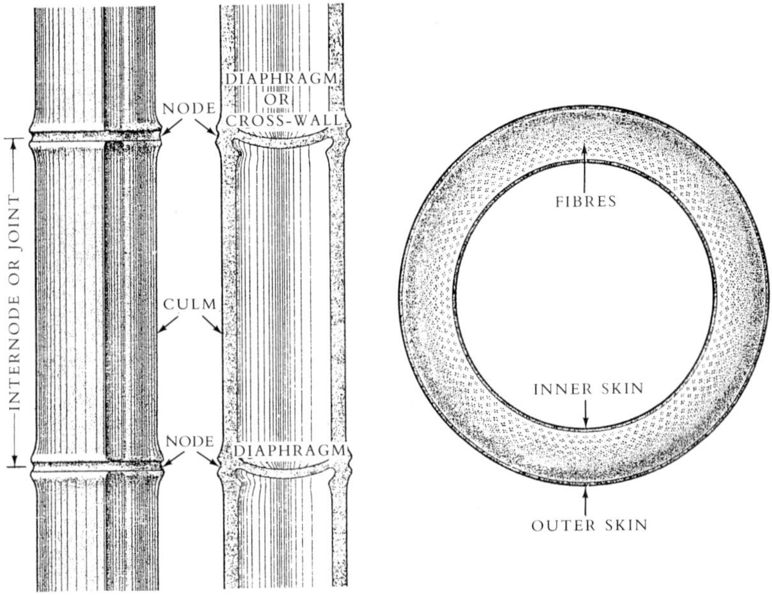

Fig. 1 Structure of a bamboo stem or culm, showing nodes, internode, and cross-wall.

The resemblance that de la Loubère saw to reeds probably derives from the clumps in which bamboo grows. In fact, reeds have unsegmented, solid, triangular stems, so are structurally quite different to the bamboos.

The majority of South-East Asian bamboos grow as tall clumps of erect stems with feathery foliage growing from the upper third to upper half of the stem. Other smaller bamboos exist and there are also climbing species which use trees for their support. Some bamboos grow densely, as a clump, such as that shown in the nineteenth-century engraving in Plate 1. Other bamboos throw up stems at some distance from each other. This type is illustrated by the bamboo grove of Colour Plate 1 which surrounds a temple.[1]

Rattan is a palm and is only superficially similar to bamboo since rattans are solid canes. A rattan cane grows only at the tip of the stem from what is called the palm 'cabbage' or 'palm heart'. In most cases, many stems of rattan (perhaps as many as fifty at a time) grow from the same set of roots. Externally, rattans appear to be jointed like bamboos as there are rings at intervals along the stem, but rattan nodes merely mark the base of a leaf sheath that once protected the stem.

Rattans are not erect and self-supporting. Instead, they climb or scramble through other vegetation, relying on hooks to get them into the forest canopy and close enough to the light.

Another early European traveller, Johan Nieuhof, remarked of Javanese *rotang* or rattans that the stem 'grows so fast and twists its self sometimes so closely up to the very tops of the highest trees, that they die for want of air'. In Plate 2 and Colour Plate 2, rattan can be seen climbing other forest vegetation. Rattans can be far longer than bamboos are tall: a 180 m rattan stem has been recorded and 120 m is not uncommon.

As rattans are solid canes they can be bent and set into curves without failing. Bamboo, on the other hand, is stiff and resilient, bending so far and no further before it fails. If not stressed too much, a bamboo cane will return to its original shape and

[1]See Chapter 7 for more information on clumping and 'runner' bamboos.

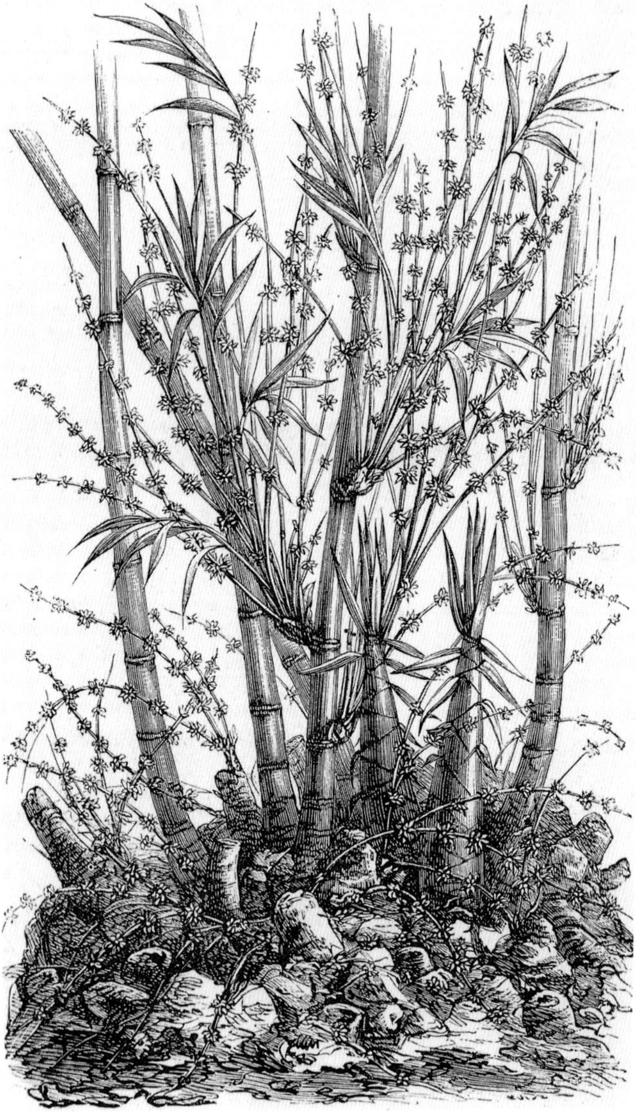

1. Mature clump of bamboo (from A. and C. Rivière, *Les Bambous, végétation, culture, multiplication*, Paris, Société d'Acclimatation, 1878).

2. 'Vegetation of Celebes' (from W. H. Adams, *The Eastern Archipelago*, London, T. Nelson and Sons, 1880).

straightness. The hollow centre contains a woolly material which can easily be removed. Once this has been done, each internode is a natural storage vessel for goods of all kinds.

By no means are all the traditional applications of bamboo and rattan mentioned in this book. Although bamboo is already exploited in so very many ways, new applications continue to be found. For instance, carbonized filaments from bamboo fibre have been found to be well-suited to special purpose electric lamps, and Edison himself used bamboo filaments. In the early days of the gramophone, bamboo was used for needles. Bamboo charcoal is reputedly preferred by goldsmiths for heating precious metals. Fish oil can be deodorized with bamboo leaves. In India and in several countries of South-East Asia the major use of bamboo is for making paper. In medicine, bamboo shoots produce enzymes of varying uses to man. The shoot is also used to produce an extract with water which may be used to culture pathogenic bacteria. Bamboo joints are well-suited to making inexpensive artificial limbs.

Although bamboo and rattan have often been replaced by plastics and other materials, their use will surely continue as strongly as ever because they are uniquely satisfying to look at, to decorate, and to use. In Western countries articles made of cane can command prices that would astonish their original creators.

The chapters that follow outline the great range of uses that bamboos and rattans fulfil in their native region: uses within the home or outside it to provide food or security, entertainment, trade, and contact with spirits. Where it is used for practical purposes, a cane will be selected because of particular physical characteristics, such as strength or lightness, which make it more appropriate than other materials such as timber or palm fibres. Where bamboo is used for ritual functions, it is more difficult to explain the repeated use of bamboo rather than the other plant materials available in the region—until one remembers how closely and frequently man and bamboo are associated in the proverbs and beliefs of South-East Asia. Some of these are quoted in Chapter 6.

2
Canes for Everyday Needs

Bamboos and rattans are put to hundreds of different uses because they are strong, light, smooth, straight, and round. Moreover, they can be split lengthwise into straight strips or fibres and they come in many different sizes with varying lengths of joint. They can be perforated, they have hard outer skins, and they are free from any pronounced taste or smell. In addition, both bamboo and rattan are found abundantly and grow and increase rapidly.

To this list of useful properties must be added the hollowness of bamboo, which makes possible another set of uses. Where other materials are used for the same tasks—be it wood, textiles, metal, or ceramic—far more labour and preparation is required. As the Victorian naturalist, Arthur Russel Wallace, said, 'The Bamboo is one of the most wonderful and most beautiful productions of the tropics, and one of nature's most valuable gifts.'

Food Preparation

Rice is, of course, the staple food of Asia and in much of Malaysia and Thailand bamboo is the preferred container for cooking rice. Unlike some cooking pots, it does not spoil the flavour of rice. To steam rice, a bamboo stem is cut into 1 m lengths then filled with rice and sometimes small pieces of meat, together with a nicely calculated amount of water. The bamboo is then placed over the fire (but not in contact with it) and a bundle of leaves is used as a lid. When the bamboo begins to scorch, the food is cooked. *Khaaw laam* is a Thai hill tribe dish, consisting of sticky rice, coconut milk, rice water, and beans, cooked in this way. Cooking tubes of bamboo are shown in No. 6 of Colour Plate 3, whilst Plate 3 shows sticky rice being cooked in bamboo tubes in Malaysia

3. Cooking *lemang* (sticky rice in bamboo tubes), Malaysia.

to produce *lemang*. The bamboo is lined with banana leaf, then glutinous rice is fed in before the tube is topped up with coconut milk. The rice is cooked slowly over an open fire for several hours.

Many other implements connected with the preparation of food are shown in Colour Plates 3 and 4. There are two steamers in Colour Plate 3 (Nos. 7 and 8), several forms of basket for holding rice, and a sieve for preparing sago (No. 13). Sieves are also used for straining coconut milk since bamboo does not taint the milk, a very important advantage. Other kitchen utensils shown in Colour Plate 3 are also made of bamboo. The conical basket (No. 9) and the decorated basket (No. 10) are for rice, the latter being used for more formal occasions. Plate 4 shows a novel use of rattan: coconut is being grated against the hard spines present on the inside of the leaf stalk or petiole.

Because bamboo resists attack by vinegar, it is fashioned into hoops for wooden barrels used for pickling fish and vegetables. In Japan, wine is kept in pots of green bamboo in order to improve its flavour.

The cooking fire itself might include some bamboo, and the fire could have been kindled from bamboo shavings such as are used in buffalo horn fire-makers by Thai hill tribes like the Lahu. The centre of stems which have died but are not yet rotten is always dry, even in the rainy season, so such stems are a reliable source of kindling. Until matches became the simpler alternative, bamboo fire pistons were commonly used. The principle of these is that of heating air by compressing it. This is done by forcing a tightly fitting plunger down a bamboo tube. The hot, dry tinder inside lights easily from a spark. In Plate 5, blacksmiths in Sarawak are using large joints of bamboo as a bellows.

To serve food, joints of bamboo are split in half to make a trough-shaped serving dish for curries, whilst bamboo may also be carved into cups and spoons. Split and woven, bamboo is used for plates, and also to make the low round tables found in rural homes. Everywhere in South-East Asia, skewers of

4. Coconut grater made from the leaf stalk (petiole) of rattan (John Dransfield).

bamboo are used for satay, or if the meat is in largish pieces, it may be cooked on a bamboo spit.

Canes for House-building

Together with reeds, bamboos are some of the oldest and most widespread building materials used in settlements across the tropics. Their advantages are many: they are cheap and readily

5. Blacksmiths with bamboo bellows (Hedda Morrison).

available, and although individual parts may not be very durable, they are easily replaced. A house built of canes is, in fact, strong but not rigid. Its flexibility also means that it can withstand earthquakes. A traditional rural house, raised on piles, will consist of posts of de-barked tree-trunks upon which are set the house frame and roof frame of bamboo under a roof of either thatch or shingles (Colour Plate 5). The lightness of the bamboo frame of the house means that whole sections, such as a roof frame or wall frame, with bamboo edges and diagonal strengthening struts of bamboo, can be prefabricated on the ground then lifted as a single piece on to the floor piles by a team of three or four men. The traditional Torajan roof shown in Colour Plate 6 is an altogether sturdier and more durable structure.

Where vertical poles are made of bamboo, care is taken to ensure that these are in the same position as when growing, and not 'upside down'. It is said that if the poles were to be upended, the family would quarrel endlessly.

House walls are often panels of interwoven strips of bamboo, though other fibres are also used. Bamboo canes are split, dried in the sun, then woven together, often in a pattern that uses the contrasting colours of the inner and outer faces of the bamboo—dark brown and light brown—to provide decoration.

A thatched roof will be supported with a bamboo ridge-pole and battens under the thatch. An alternative roofing material is bamboo split in half along its length, then pinned to the roof frame. Bamboo shingles or 'tiles'—rectangles of bamboo tied to a frame with rattan—are also used. Increasingly, other materials are used in house-building. They may be more modern but they are not necessarily better. Palm leaf thatch, for instance, offers far better insulation from the sun than does corrugated iron.

The floor is very likely to be of bamboo, split then flattened to give narrow planks or laths of smooth surfaced, springy flooring which is pleasant to walk upon (Plate 6). In Thailand, a traditional Hmong birth is said to have entailed the newborn baby dropping on to the bamboo floor from the birthing stool

6. Weaving rattan, and house interior (Hedda Morrison).

on which the mother sat, supported by her husband. The springiness of the floor cushioned the baby's fall and, say the Hmong, encouraged the child to start breathing.

In some cases, two or three layers of split bamboo will be used for the floor, and mats, also often of bamboo, may be laid on top of this. There is a non-woven matting called 'smooth matting' made in China and further south. Sections of culm are selected from the middle part of bamboo stems, below the branches, and the nodes planed or sanded flat. The leaves of the *loguitip* tree may be used for this in the Philippines, or *daun amplas* in Malaysia; both are natural 'sandpapers'. The culm is split into 2.5 cm strips of bamboo which are flattened then drilled through and strung together with strong cotton to make a light and comfortable mat for covering beds and cots. The sleeper lies on top of the smooth outer layer of the bamboo. The bed is said to be particularly comfortable in the hot, sticky tropical night, and to improve with use. Even the pillow may be a simple bamboo joint.

Bamboo offers a solution to the seasonal flooding common in parts of Thailand. An ingenious solution is to build houses over a raft. As the waters rise in the wet season, the houses rise sedately, too, lifted up by the bamboo poles stacked beneath the floor (Plate 7).

The simple furniture of the rural dwelling, as well as fashionable city furniture, may be made of bamboo and rattan. Low circular stools, waisted or cylindrical, are common across the region and are usually made from a combination of rattan and bamboo. The main weight-bearing members are of bamboo culms or hardwood, the vertical struts may be of rattan, and the woven seat section of bamboo or rattan. Low circular tables are made in a similar way, frequently with a raised lip made from a length of rattan encircling the edge and lashed to it with split rattan. Plate 8 shows rattan furniture—the major use of rattan in South-East Asia.

Rattan also has its place in house-building, often as a means of lashing posts together, or for weaving into wall frames. Rattan is also used for matting. In Sarawak the Penan people

7. Floating houses in Thailand (Sumet Jumsai).

8. Cane furniture from Thailand (Brian Pope).

are famous for mats woven with the finest rattan and there is also a type of mat that is very similar to the 'smooth mat' made of bamboo. This is the *tikai lampit*, a large mat of up to 2.5 m × 6 m made of split rattan canes approximately 1 cm thick, laid side by side, pierced and threaded together. In order to make a secure edging, the end of the canes are crushed to break them down to fibres, and then plaited into a decorative border. Mats are just one of the many products available at the shop shown in Plate 9, which also offers hats and baskets of bamboo, rattan, and other fibres.

Finally, whilst bamboo and rattan seem to have been replaced for house-building in cities in modern times, the two species are still hard at work in the construction even of very tall buildings, serving as a scaffolding that is light but strong, easily erected, dismantled, and transported. These advantages are also important for travelling theatres which need to be erected and taken apart rapidly and transported easily. The 'theatre' consists mainly of a wooden stage on bamboo beams lashed

9. Mats, baskets, and hats for sale in Sarawak (Hedda Morrison).

together. The walls are bamboo mats, brightly painted and traditionally decorated. There may be a highly decorative roof with ornamental symbols. The whole is supported with bamboo braces and, if necessary, guy ropes keep it steady.

Canes in Agriculture

The uses of bamboo in agriculture are many: rakes, handles for hoes, irrigation wheels and pipes, racks for drying rice, and rice scoops. Living hedges of thorny varieties of bamboo may protect livestock, farm houses, or a whole village. Bamboo screens protect orchids from the sun and vegetable crops from wind or rain. Bamboo poles are used to support fruit trees and to enable red ants to bridge the gap between citrus trees in an orchard. The red ants cross to prey on the sap-sucking scale insects which infest the trees.

Land must be ploughed and produce must be sold at market, and bamboo comes in here too. South-East Asian buffalo used for draught are fortunate when their yokes and carts are made of light bamboo. Where a cart is an unattainable luxury, goods will be taken to market in baskets suspended from a springy bamboo pole, like the Javanese carrying pole shown in Colour Plate 4, No. 7.

Scarecrows are made from bamboo, for example, the Balinese or Malaysian wind-wheel atop a pole. It rattles noisily in the slightest breeze to scare birds. The Iban manufacture a bamboo bow with a plaited bark bowstring and hang it from a high pole. A hardwood clapper is fastened in the centre of the bowstring, and when the breeze blows against a sail made of palm fronds the clapper beats and cacophony ensues. Another, quieter, type of bird scarer is made of the leaf sheaths of bamboo. A bamboo pole is set in the ground at an angle so that a leaf sheath suspended from a short cord can hang free. The outer face of the sheath is dull but the inner face is pale and polished and flashes in the light when blown by the breeze.

But scarecrows can fail and life sometimes offers difficult choices. A Malay proverb warns: 'Deaf sparrows are eating your rice out in the rain. Drive them away and you will get wet; leave them and the rice will be finished.'

Construction

Bamboo and rattan are also used extensively outside the home for minor and even major works of civil engineering. Wallace tells of a use of bamboo for climbing trees that he saw in Borneo when collecting zoological specimens. On arriving in a remote part of Borneo, Wallace deemed himself very lucky to have shot a *mias-kassir* (a small orang-utan) on his first day. The *mias-kassir* fell from its branch but remained fixed in a fork of the tree. Wallace was anxious to get it and asked the Dyaks with him to cut down the tree. To his surprise they said they would prefer to climb the tree, a task he believed impossible as the first branch was almost 20 m above the ground. The Dyaks, however, cut a stem of bamboo from a nearby clump which they converted into a set of two dozen pegs, each about 30 cm long and sharp at one end. Cutting a piece of wood for a mallet, they then drove pegs into the tree at intervals of about 1 m, and tested them.

Wallace looked on with interest, convinced that their method was flawed, for if any peg should fail the climber would fall, and if the peg were at a good height the fall could be fatal. Then another bamboo stem was selected, cut and stood next to the tree and its pegs. The long stem was tied firmly to the pegs by a cord made from bark. As the climber stood on the first peg, he drove in another higher peg. This was then also bound to the vertical bamboo and the man climbed up another rung. So he continued until it was necessary to haul up another bamboo stem, tie that to a few of the pegs already firmly attached to the lower bamboo stem, and then continue upwards. In this way, the climber's weight was distributed between several pegs by virtue of the vertical bamboo stems (if any peg failed, the man would still be secure on the remainder) and the *mias-kassir* extricated from the tree. These bamboo ladders are used by the Dyaks to gain access (sometimes at night) to combs of honey and wax which bees build suspended from the branches of the loftiest of forest trees. The same technique has also been observed in Myanmar (Burma).

Plate 10 shows bamboo used for water supply, both to deliver a shower when halved and to collect water to be carried back to the house. The Lisu of Thailand prefer to site their villages over the ridge from a stream so that they can then divert water for their needs through bamboo aqueducts, and at the same time reside at a safe distance from the spirit powers that they associate with streams and water.

Bamboo has uses in communications. Steps could be fashioned from bamboo up steep hillsides: 1 m-long lengths of bamboo were notched at either end, laid in place, then fixed with pegs hammered into the ground through the notches. Whilst such steps will decay within 6–12 months, the ease with which they are built makes bamboo more economical to use than a more durable timber.

Wallace also refers to bamboo bridges across streams supported in the canopy of tall forest trees growing at the water's edge. Bamboo bridges also led up to granaries built high up in trees overhanging the river. Another type of bridge was seen by Wallace in Dyak country where gullies and ravines separated villages. A series of X frames were made with pairs of stout bamboos. The frames were set upright in the ground and lashed firmly together with the upper arms rising a few feet up from the crossing point. A large bamboo was then laid along the series of X frames and this constituted the pathway. Another bamboo, higher up, served as a 'slender and shaky' handrail. Similar bridges and footpaths over marshy ground can be seen today in Sarawak.

Many types of bridge are built across the rivers of South-East Asia, from the most simple—a pair of rattan lines, one to walk on, one to hold—to the most elaborate bamboo and rattan suspension bridges. The Min River in Szechuan Province, China, is spanned by a bamboo bridge 225 m long built entirely from bamboo. The original idea for suspension bridges probably comes from Eastern Asia and bamboo would have been the material used.

Another bridge may also be mentioned. For the Burmese

10. Washing under a bamboo shower, and collecting water in bamboo vessels (Hedda Morrison).

Kachin people, the bridge of death is a bridge made of a single bamboo stem. It stretches across serried ranks of cauldrons in which noisome liquids simmer and bubble. As the wicked cross over the bridge, the contents of the cauldrons boil up to engulf the wrongdoers.

Ornamental Use

Whilst the bamboo is most widely grown for very practical purposes, it is also grown for purely aesthetic reasons, as an ornamental plant in gardens and parks. Bamboo groves are traditionally locations for lovers to meet or dream:

I sit alone in the dark bamboos
Play my lute and sing and sing
Deep in the woods, where no one knows I am
But the bright moon comes and shines on me there.

(Wang Wei, AD 699–761, translated by G. W. Robinson)

As an ornamental plant, bamboo combines a slender arching stem with feathery foliage which rustles attractively in the breeze. Stems of live bamboo are very varied: there are black, yellow, green, striped, and spotted species. The culms are of different shapes, most being round but some bulbous or curiously twisted. The distance between the nodes can vary from 3–4 cm to 1 m or more. In some species the internodes are straight and the node is merely a slim and narrow ring around it. In others, the internode bulges interestingly, as in Plate 11 which shows *Phyllostachis aurea*. Like other bamboos, this has a groove in each internode, above the point where the branches emerge from the stem.

There are so many bamboos, they cannot all be mentioned and not all are suited to all environments. Some are definitely tropical while others can survive cooler conditions. Some particularly attractive species are:

Buddha bamboo (*Bambusa ventricosa*): This species occurs in dwarf forms and up to 5 m tall, though a clump will normally include culms of different sizes. The internodes are short and attractively pear-shaped. Bottle green when young, the culm gradually turns yellow with age. The leaf sheaths are also green at first but turn from orange to the colour of straw and then fall off. This is a species which has not been observed in flower as it always reproduces vegetatively.

Golden bamboo (*Bambusa vulgaris var. striata*): This is a large

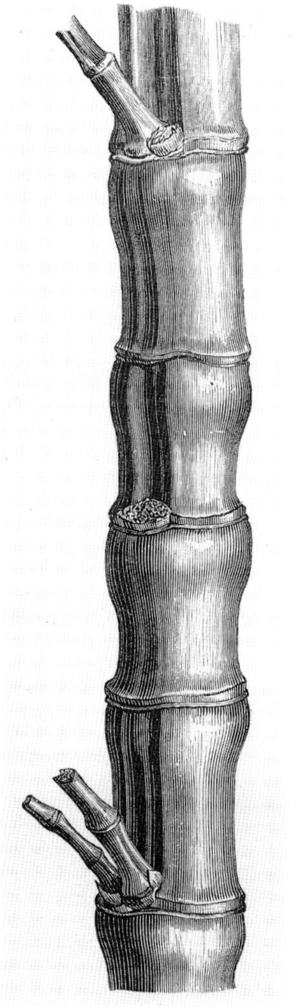

11. Bamboo stem of *Phyllostachis aurea* (from A. and C. Rivière, *Les Bambous, végétation, culture, multiplication*, Paris, Société d'Acclimatation, 1878).

bamboo growing to 12 m with stems about 8 cm in diameter, coloured yellow, and striped with green. The internode is about 15–20 cm long and the nodes are prominent.

Punting pole bamboo (*B. tuldoides*): Attaining about the same height as golden bamboo, the stems of this species are a little thicker, up to about 12 cm in diameter. The colour of the stem is a delightful bluish-green which has a white bloom initially. The stems become yellower as they age. Punting pole bamboo can be identified by its colour and the smoothness and total lack of hairs on the leaf sheath. Most bamboo species have a dense coat of dark, sometimes reddish, hairs on the sheath surface.

Black bamboo (*Phyllostachys nigra*): A bamboo with a purplish-black stem.

Tortoise-shell bamboo (*Phyllostachys edulis var. heterocycla*): This is so called because of the unusual shape of the joints. These are short and the line of the node is not regular and horizontal but rather rises diagonally to alternate sides, giving a zigzag across the stem. The joints thus bulge like the shell of the tortoise. Square bamboos may also be seen, but this is achieved by tying four flat laths to the bamboo stem and forcing it to grow into this shape.

Some bamboos have a slightly phosphorescent greyish-white bloom on the surface of the young stem. On some stems this is faint, rather like the bloom on grapes or plums, but on some species it is thicker and almost fluffy or flour-like.

In calm loneliness I shut my door
Against the whole afterglowing sky
Cranes are nesting in all the pines
No visitors at my wicket gate
Tender bamboos with the new bloom on them
Red lotuses free of their old garments
A lamp shines out at the ford
Water-chestnut pickers come home.

(Wang Wei, AD 699–761, translated by G. W. Robinson)

As a garden plant, bamboo does have the disadvantage of raining down a great quantity of leaves which are slow to decompose and which must be removed before they clog up water courses and drains. Bamboos of the 'clumping' kind are clearly preferable in gardens as they do not spread widely (see Plate 1 and Chapter 7).

Apart from their beauty, another purpose of planting bamboo can be to conserve soil since the rhizome and its roots are very effective in protecting slopes against destruction and removal by flood waters. Bamboo mats have also been used as a temporary measure to protect the banks of rivers and lakes from flood water.

3
Food, Medicine, and Magic

Food

My new province is a land of bamboo groves
Their shoots in spring fill the valleys and hills
The mountain woodman cuts an armful of them
And brings them down to sell at the early market.

Things are cheap in proportion as they are common
For two farthings I buy a whole bundle
I put the shoots in a great earthen pot
And steam them over the boiling rice.
Their purple nodules broken, they suggest old brocade
Their white skin gleams like new pearls.

(Po Chu-i, translated by A. Waley)

DURING the rainy season, which is when the new buds emerge and shoot, bamboo constitutes a large proportion of the diet of some tribal peoples of the South-East Asian mainland. When the bamboo shoot pushes up through the ground, it is filled with sugars to enable it to grow very rapidly to join its fellows in the canopy. This food is attractive to wild animals and to man, but it is surrounded by spines and prickles and by layers of leaf sheaths ready to grow to protect the young stem. Some bamboo shoots are also protected by the toxic compounds they contain, although these toxins can be removed by cooking. Some tender varieties require only ten minutes boiling for their preparation, but for tougher varieties it is necessary to boil the shoots for an hour after trimming, to discard the water, then boil again in fresh water for a further hour. The rather fibrous shoot is then eaten as a cooked vegetable or cold in salads (Colour Plate 7). Bamboo shoots can be preserved either by drying or pickling in salt water or the water used to wash rice. The young shoots of a bamboo from West Java (*Gigantachloa apus*) are bitter but are edible after being soaked in mud for 3–4 days.

As a member of the grass family, together with wheat and rice, bamboos also produce grain. This is not commonly used as food as it is not a regular or reliable crop, but bamboo grain has often been a 'famine crop', and it has been noted in India that years of poor crops may see abundant seeding by bamboos. This may be merely fortuitous or there may be a causal link. The seeds are usually small and are harvested by birds or rodents. The seeds of one bamboo (*Melocanna baccifera*) are larger, about the size of a small pear. These seeds germinate very early, while still attached to the cane, putting out a root before falling to the ground. This early germination has presumably evolved as a precaution against the depredations of wild animals which would otherwise consume too much seed. Bamboo seeds are probably of far greater importance to wild animals than to man, but a corollary of this is worth noting: where a large clump has flowered and produced grain, it will encourage the build-up of large numbers of rats and other pests. Once the grain has all been eaten, these animals will turn their attention to any nearby cropland.

The leaves of bamboo are used as fodder for livestock and are as nutritious as more common fodder grasses. Elephants are very partial to bamboo leaves, and further north the panda depends on the bamboos of cooler climates, eating the shoots.

Bamboo is the home of another type of food. Caterpillars that live in bamboo shoots are considered a delicacy in some places: *kabue* and *beetung* are two caterpillars toasted and eaten in northern Thailand.

Rattans produce fruit more regularly than bamboo and so are a more reliable source of food for animals. Their fruits are covered with hard, shiny scales which overlap in neat patterns, each with a groove from base to pointed tip along the mid-line. The scales are so ordered that the groove of one scale leads to the groove of the next one, making vertical lines along the fruit (Colour Plate 8). Rattan fruits may occasionally be seen in markets in remote areas. They are similar to the fruits of two more common palms, the zalacca palm and the sago palm. All three belong to the 'scale nut' subfamily of palms (*Lepido-caryoideae*).

The scales of rattan fruit vary from yellowish brown to black, with some striking exceptions, such as a magenta-coloured fruit and an ivory and black particoloured fruit. The brilliant red of some *Daemonorops* species comes from a red resin which thickly encrusts the scales. This is known as dragon's blood and is used in various ways in handicrafts and medicine (see Colour Plate 8). Within the scales of rattan fruits is a layer of flesh which may be sweet and tasty or sour and inedible. Inside this flesh lie the seeds. In the Philippines, for example, *limuran* (*Calamus ornatus*) is sour or semi-sweet, and the juice is sucked out. *Litoko*, a related species from northern Luzon, is eaten whole and may be pickled.

The tip of the rattan stem, the palm cabbage or palm heart, is enjoyed by wild animals in the forest and also by man. However, whereas harvesting young bamboo shoots does not greatly disturb the overall plant (that is, the bamboo clump), collecting a rattan heart kills a valuable palm stem. The rattan heart is, nevertheless, toasted by Thai hill tribe peoples, and eaten as a slightly bitter but aromatic vegetable. It is also eaten curried.

Medicine, Magic, and Poison

Stem oil and sap, roots, leaves, and mineral accumulations in bamboo are all used as medicines. The oil in bamboo stems is claimed to have curative properties. To take advantage of these, eggs are cooked slowly in a split joint. The eggs absorb the essential properties of the oil and this purportedly enables them to cure maladies as diverse as asthma and consumption. In Malaysia, bamboo was once used in poultices whilst the shoots of various species were used to treat gonorrhoea and kidney trouble. In Cambodia, the nodes are boiled to make medicines for chest complaints and urino-genital disorders.

The idea of signatures is one that flows through local medicines. This is the belief that a physical property of a plant indicates what it may be used to treat. The colour yellow, for example, is associated with epidemics, so *Bambusa vulgaris*, a yellowish, striped or spotted bamboo, is exhibited on a tray to help exorcise epidemics. This bamboo also serves, for the same reason, in the treatment of jaundice. The watery sap the stem contains is esteemed as a remedy for wasting diseases, while boiling the roots gives a remedy said to relieve kidney trouble. The stem sap of a scrambler bamboo of Luzon (*Dinachloa scandens*) is used as an eye medicine, and its young shoots act as a vermifuge.

Also in the Philippines, *Bambusa spinosa* is a multi-purpose medicine: its leaves are used as an anthelmintic, its roots for spots and sores, and the young shoots are used as a poultice for dislodging worms from ulcers.

The hard white substance sometimes found in bamboo stems, tabashir, is the object of various beliefs. Called *kapur buloh* (bamboo camphor) in Malaysia and *tahi bambu* in Java, tabashir has all the requisites of a product about which semi-magical ideas can gather: it is found in the dramatic circumstances of a burned-over bamboo grove and is known to accumulate within the bamboo stem. Concretions found within plants are known as bezoar stones and are often accorded great powers and are used in faith-healing in Malaysia. (Hard lumps of resin found inside jackfruit wood are also considered to have magical powers.)

Tabashir is now known to be composed mainly of silica, a mineral which is also accumulated throughout the wood of older bamboo stems, improving their hardness. As a stone it is white and pearly when wet, opaque when dry, but it is often sold blackened by the ash of the grove from which it was collected. The range of ailments it is said to treat makes it a panacea for coughs, consumption, asthma, fever, and illnesses of the lungs generally. It is, moreover, said to act as a tonic and an aphrodisiac. Several grades of tabashir are recognized in traditional markets for the stone. A form of tabashir with a sugar content high enough to make it sweet has also been described; this was claimed to 'cool the heart, the brain, the liver or the kidneys', as well as checking diarrhoea. Tabashir became known in the West at an early date, being available in the markets of Alexandria in classical times. Although the Greeks did not make much use of it, it was mentioned by Arabian physicians such as Avicenna from AD 940, and Chinese medical authors also mention it around this time. A travelling friar, Odoric, who visited Java at the beginning of the fourteenth century, quoted a story that claimed that the stone would act as an amulet, making its owner invulnerable if embedded in his body.

Another bamboo product sold as an aphrodisiac, though probably even less effective, is the rhizome of the bamboo *Dendrocalamus hamiltonii*. With a little trimming and dressing, this is made indistinguishable from rhinoceros horn, which commands very high prices in mainland Asia and Japan.

Bamboo's properties are not all curative, however, and there is evidence that it may also be put to criminal use. The sheaths that surround the stem are covered with brown or red protective hairs which have sharp tips capable of penetrating the skin. If an intended victim can be induced to swallow these hairs mixed into food, the resultant irritation and poisoning of the intestine can kill. A recipe for an abortient is based on a shrubby bamboo (*B. multiplex*). The shoot is pounded with a hundred white peppercorns to produce an infusion which should be drunk on three successive mornings to obtain the desired result.

But bamboo is not so frequently associated with evil and sadness as it is with youth, flexibility, and love. There is a bamboo known as 'bamboo for passionate yearning' in Malay, which grows isolated in mountains of the Malaysian peninsula. It is sought after by the lovelorn to inflame the passions of a loved one. Another bamboo with a similar name was used by theatrical troupes to enslave their audiences to their will. It was said that a thin sliver of 'yearning bamboo' inserted between the teeth could enchant the listener. Belief in this power was once such that the mere possession of this bamboo was a punishable offence in the Malaysian peninsula.

Rattan products have also stirred the imagination. Dragon's

blood, the resin which oozes from between the scales of the rattan fruit, is used as a medicine. Though no longer of much importance, it was once a major commodity of trade between South-East Asia and China. When the fruits are shaken in a basket filled with cockle shells, the resin is scraped off. It then falls out through small holes in the basket as a gritty powder. This powder is pounded, then softened with hot water and kneeded into lumps and sold as an ingredient of local and Chinese medicines, often as an eye medicine. In Europe, it became known as a cure for dysentery and diarrhoea.

This ritual use of dragon's blood comes from the Semai people of Peninsular Malaysia: a menstruating woman must take care not to approach 'tiger rattan' (*Daemonorops didymophylla*) too closely. If she does, a fruit may fall and grow into a cat and then into a tiger which will follow her everywhere. The tiger is exorcised by a gift of dragon's blood kneaded into fruit-size balls and left on a mountain top after magical spells have been whispered over a shaman's fire.

The Philippine rattan species *limuran* (*Calamus ornatus*) produces sour but popular fruits. Herbal medicines recommend the ashes of this species for treating chronic yaws and also as an anaesthetic in childbirth. Amongst the Semai, the leaves of this rattan are bundled with leaves of other 'magical' plants and waved over all members of the community when an epidemic erupts. The roots of rattan are said to be used in Cambodia to produce a decoction taken at meal-times as a tonic for a variety of complaints of the digestive tract and to reduce fever. It is also used in veterinary medicine. The roots of *rotan tawar*, or water rattan, are infused to made medicine for elephants, though it is perhaps too powerful a drug for man: unconsciousness results if carp gall is mixed with the juice of *rotan tawar* and taken internally. Another poison is derived from rattan by scraping the glassy surface from the rattan stem. Although this silica deposit is chemically the same as tabashir, in this form it is said to act as an irritant to the gut—especially when mixed with powdered glass!

4
Handicrafts

READILY available and inexpensive, the canes play a role in all sorts of handicrafts, either as the tools used or as the crafted articles themselves. Baskets, mats, hats, and boxes of various kinds are made using canes because they are light, easy to work, strong, and resilient.

Baskets

Both bamboo and rattan are fissile, that is, they cleave easily from top to bottom, though bamboo, in particular, is hard to cut across. Cane stems are tender when they first shoot, but over two or three years lignin is laid down in the cells and wood is formed. Later, silica accumulates in the outer layers. Where hardness and durability are needed, therefore, older stems are employed, but where flexibility is important, younger stems, perhaps only one or two years old, are used. Baskets of all kinds and for all purposes are made from bamboo and rattan though the canes are soaked prior to weaving to make them more supple.

The variety of shapes and uses of baskets is huge: wherever a container is needed, one can be made from cane. Colour Plate 9 shows a group of Thai 'lunch-boxes' waiting to be sold. These will carry cooked rice and are made of woven bamboo, with a wooden base. There are oval horse-panniers made by the Lisu and Meo in northern Thailand, there are cylindrical bottle-carriers (see No. 8, Plate 12), and there are rice bowls and spirit cups made from bamboo and rattan. Clothes are also stored in bamboo bins, as is rice. Cane baskets are typically left to 'cure' in the smoke from the cooking fire. This changes their colour to a honey brown and also enables them to withstand insect attack.

Open-weave baskets are made for goods like large vegetables

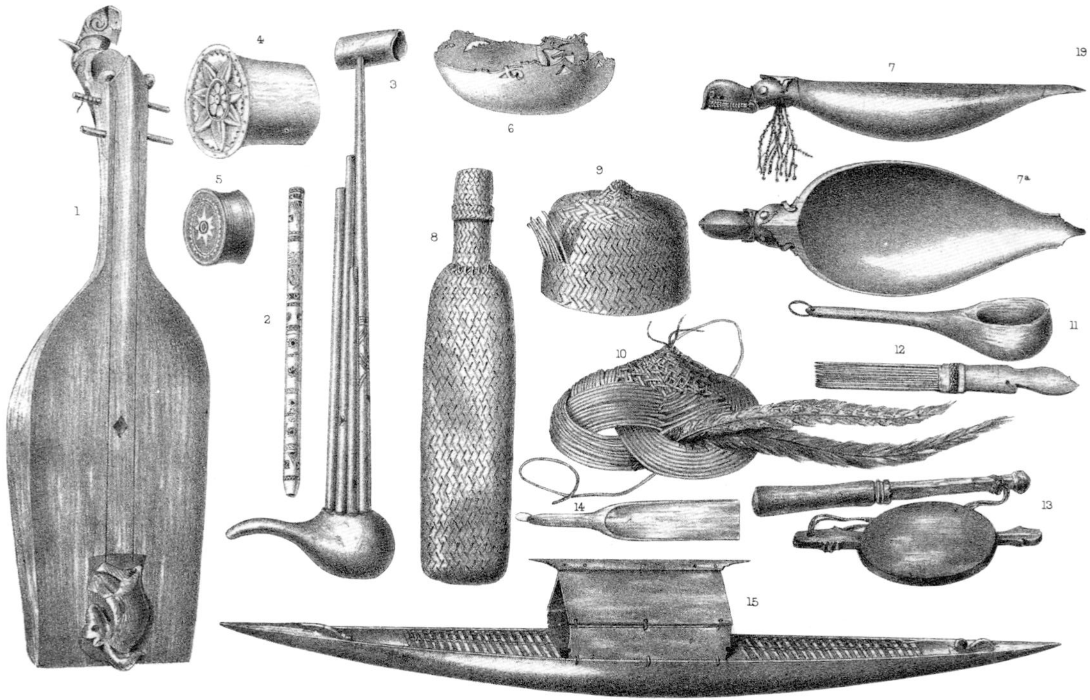

12. Dyak utensils and canoe (from C. Bock, *The Head-Hunters of Borneo*, London, 1881).

13. Open-weave baskets for transporting cocks (Brian Pope).

or even animals (Plate 13), and where river banks need stabilizing or a stream needs to be forded or dammed, stones are packed into bamboo baskets called gabions to hold back flood or landslip. Close-woven containers, which may be treated to make them watertight, are made for liquids and powders. Baskets are also used for carrying seeds at sowing time (rice, cotton) and for collecting the harvest (tufts of cotton, ears of rice). Such baskets are often woven diagonally, starting with the centre of the base. Once the base is large enough, the weaver starts to bring the sides up by 'turning a corner' and weaving together canes from adjacent sides of the base. The sides of the basket, the base, and the lip are often strengthened with rattan or other rigid, wooden reinforcement.

Trays of split and woven bamboo are used for presenting goods at market, for drying produce in the sun, and for winnowing rice (Plate 14). Silkworms hatch and live their lives as caterpillars eating mulberry leaves spread out across bamboo trays. When they pupate they do so in bamboo racks

14. Rice winnowing trays from Perak, Malaysia.

designed to resemble the tree tops that are their natural habitat. Colour Plate 10 shows a collection of cane bird-cages for sale in a Thai market.

An open-weave basket may be lined with another fibre, such as palm leaf, to give another texture and another set of properties: they are lighter, easier, and faster to make but closed to the atmosphere and proof against some insects. Woven bamboo casks, for example, are lined with broad, stiff bamboo leaf sheaths so that they may be used to carry incense powder. For close-woven baskets, a 'putty' made of crushed beans or similar material may be pressed into the interstices before the baskets dry out and harden, to make them watertight. Beeswax and lacquer are also applied for the same purpose. The Indonesian gunpowder boxes shown in Colour Plate 11 are made mainly of lacquered rattan. Colour Plate 12 shows three rattan baskets used by the people of West Kalimantan to carry pepper over the border for sale in Sarawak.

There are several basic types of basket weave which may be

15. Plaited rattan backpack (Hedda Morrison).

combined in a single object. Instead of the simplest warp and weft, one-over and one-under, of split canes, the weaver may prefer to use several-under and one-over or some other combination. She may set the warp and weft at an angle to produce a herringbone effect, or she may use the twining technique which requires that the weft canes cross one another at the back of the work in between two warp canes. The result is that only the weft canes are visible and the warp is hidden. Colour Plate 13 shows some delicate weaving styles used on simple Thai handicrafts.

Coiled basketwork uses a thick strand coiled in a tight, flat spiral which is kept in shape by sewing it together with string or by joining adjacent coils by wrapping another softened fibre around them. The vertical sides of a coiled basket are built up in a similar way. The edges of large baskets are usually reinforced, often with a strip of rattan. Sides and bases may also be reinforced with wood, bamboo strips, or rattan.

Plaited rattan makes elegant open plaitwork. Plate 15 shows a backpack of plaited rattan, strong enough to carry a heavy vase and leave the bearer's hands free for his trek along forest paths. The backpack is a frame and support on to which the elaborate vase, possibly a family heirloom, is tied.

Whereas textile weaving is an almost exclusively female occupation across South-East Asia, basket weaving is done by both sexes. Some division of labour may exist, though. In northern Thailand, for example, men of the Lisu tribe will traditionally prepare strips of bamboo and rattan for weaving, but it is the women who weave the actual baskets. In other hill tribes, men will also weave.

Hats

Hats are woven in similar ways, and decorated with different styles of plaiting, twining, and coiling of the fibres, which are sometimes dyed to different colours. Rattan, for example, is dyed by soaking it in water for several days or by dipping it in indigo. It is turned brown by being left to hang in smoke, or dyed red with dragon's blood—the rattan fruit resin. Burying rattan turns it black. Although alternative dyes may be used these days—onion-skins, ashes, tannins in oak and nutshells, charcoal, tea leaves, and cassava roots—the traditional colours are still used. Sometimes hats are lined with strong, impenetrable leaf sheaths from bamboo culms in order to make them both shady and waterproof.

Colour Plate 3, No. 2 shows a woven hat from Indonesia. Many South-East Asian hats are designed to shade the head whilst leaving it open to cooling breezes. To do this, a 'crown' is made of bamboo which fits the head exactly, and the hat is then fixed on top of that. The crowns are thus made for individual heads, but the hat will fit any crown.

Lacquerwork

The light and elegant lacquer boxes, trays, and ornaments produced in Thailand and Burma are often based on light baskets of fine woven bamboo. Other materials, such as reeds, grass, or wood, may also be used. An advantage of bamboo over wood is that it does not contain resin, which can interfere with the lacquering process. Bamboo is also stronger than reed or grasses. Combinations of materials are also used. For example, forms for bowls may be made of bamboo spokes which are then interwoven with reed. Once the frame has been made, several layers of lacquer are built up over it, then sanded to produce a perfectly smooth surface. Successive layers may be of different colours, then the top layers incised to reveal a pattern picked out in the colour of deeper layers. South-East Asian lacquer is traditionally made from the sap of the lac or sumac tree (*Rhus verniciflura*) which dries well in hot humid conditions but can cause skin rashes.

Pots and Boxes

Cylinders of bamboo in its natural form are used and decorated in many ways. They may be used to carry messages, to carry the shuttle of a loom, or to store precious belongings, such as jewellery, money, tobacco, or betel quids (see Colour Plate 4, Nos. 12 and 13). The base of the bamboo pot is formed by the lower cross-wall of a joint while the lid may be made either from another material or from the same bamboo stem. The outer wall of a short section of the main body and the inner wall of the lid section are removed so that the two parts can then be fitted together as the lid slides over the body wall. Pots made in this way from a joint of bamboo do have a tendency to break, so a bracing of plaited rattan is frequently added at the top and bottom.

Bamboo pots may be carved and patterns incised or scratched on the outside to reveal the paler inner stem. This may then be coloured and the incision filled by rubbing dragon's blood resin into it, or by dyeing with ink or plant dyes. The outer skin of the bamboo resists the dye but the inner wall will take it up. An example of a design is shown in Plate 16. In the Philippines, dyes are transferred from coloured paper to the bamboo which has been sandpapered to remove the resistant surface. The bamboo stem is wrapped in a layer of manila paper with a cut-out design and another layer of paper coloured with dye. The whole is then wrapped in banana leaves and boiled in water for

16. Engraved design for bamboo box for storing betel (H. Ling Roth, *The Natives of Sarawak and British North Borneo*, London, 1896).

a time. Colour bleeds through the stencil holes to the prepared bamboo surface, colouring it in the same pattern. These containers, used for pencils, dried flowers, paint brushes, etc., are known as *lokub* in the Philippines and the process (with variants) as *lakusan*. Once the dyeing process is complete, the *lokub* is polished with an oily young betel-nut.

Pokerwork is another way of decorating these pots and other articles made of bamboo. The pattern is burned on, or a pattern is cut or scratched in the surface of the bamboo, then filled and stained with soot.

Bamboo Carving

Although this craft is perhaps more widely practised in China and Japan, an account of bamboo handicrafts cannot overlook bamboo carving. The bamboo stem offers limited potential for carving as it is hollow, but functional items may be carved in relief, cutting through the surface skin to give spatial depth, light, and shade to a landscape or a plant or animal design. Alternatively, the artist cuts deeply into the culm, leaving high-relief figures, temples, or pathways in miniature. 'Wise men in a bamboo grove' is a favourite theme (Colour Plate 14). The bamboo rhizome, which is usually solid or nearly solid, is often used for carving. A Thai pipe made from a rhizome is shown in Colour Plate 15. It is set with metal and required very little carving, but in Colour Plate 16—'Boy with buffalo'—the artist shows great inventiveness in arranging the figure of the buffalo so that it could be carved from a rhizome. The boy, the buffalo, and the stand were carved from separate rhizomes. Colour Plate 17, simple but beautifully lacquered, shows a Japanese bamboo brushpot decorated with a rooster and hen.

Dating bamboo carvings is very difficult. Some are very old, but the seventeenth century is generally considered to be the 'golden age' of Chinese bamboo carving. Although carvings are often marked with the symbol for the year in which they were made, this is not conclusive given the sixty-year Chinese cycle of years. Other clues, such as the subjects handled, are also used in dating. After 100 years or so, a carving develops a patina from continued handling and polishing. Once this patina has stabilized, even experts have difficulty in differentiating between pieces that are, say, 120 or 360 years old.

Paper

One of the most important present-day uses of bamboo throughout Asia is for making paper. In fact, two-thirds of India's paper is made from bamboo, whilst Indonesia, Burma, and Vietnam also use great quantities of bamboo in this way in large paper mills. But paper-making is an ancient craft and bamboo has been traditionally used, not only for making the pulp, which is squeezed, dried, and pressed into paper, but also for the moulds in which this is done. Old bamboo stems produce a dark brown paper suitable for coarse wrapping papers. For finer papers, the upper part of the bamboo shoot must be used as here the fibres are youngest and contain least lignin.

For the finest handmade paper, the pulp is 'combed' with a bamboo loop to remove coarse fibres, stirred with a bamboo pole, then screened through a bamboo screen made from woven strips of bamboo sliced very precisely by being pulled through a hole in a piece of steel. The result is like wire: uniform in size and very fine. These fine strips, about 1 mm wide, are woven into a rectangle supported on a bamboo frame, then lacquered for durability. Semi-liquid bamboo pulp is poured over this screen, then the water is squeezed out using a press, leaving thin sheets of wet paper. These sheets are removed very carefully with bamboo forceps, then left to dry before being coated with size to improve the writing surface.

Textiles

Bamboo and rattan are used in tools for the production of textiles, as elements in the design of textiles, and as the substance of the textile itself.

Bamboo is very frequently used in the construction of looms. The backstrap loom, widely used in South-East Asia, often has a warp beam and a cloth beam (the beams between which the cloth is woven) of bamboo. A free-standing loom for larger pieces of cloth may also have bamboo stems as uprights and in the main frame of the loom.

Ikat weaving is a long slow process to produce textile pieces of great intricacy and beauty. Bamboo is used in secondary ways at different stages in the process, for example, for holding the white thread which is to be dyed by a resist process before weaving begins. Bamboo is partly chosen for this because, whether wet or dry, it will not stain the thread, as might wood or metal.

Bamboo also plays a minor role in batik work. The *tjanting* pen which is used to apply the hot wax to the fabric before it is dyed, comprises a small brass cup with a fine spout tied to a bamboo handle (see No. 4, Colour Plate 4). As in all cases where it is used as a handle, bamboo is pleasant and light to hold, smooth, and splinter free. The beeswax used for batik work in Java is taken from hives of small, harmless bees (*Melipona minuta*). To produce a domestic supply of this important commodity, a hive will be provided for swarming bees then suspended under the high roof of the house. The hive offered to the bees is often made from a single joint of bamboo, both ends closed by cross-walls into one of which a small hole is pierced for access.

Despite being strong and woody, the fibrous canes may be used to produce a variety of textiles, including clothes. Where they are used in clothing, the canes are sometimes chosen for their stiffness, as they lift other clothes away from the body. For example, the immensely ritualized Baba wedding in the Moluccas, still practised by a few traditional Chinese families, requires that the bridegroom wear several layers of clothing, each more highly decorated than the one below. An exquisite brocade jacket completes the wedding attire. Under all this, maintaining the cloth at a few millimetres from the body, and allowing air to circulate and cool the skin, is a bamboo jacket made from fine strips of woven bamboo. Several early European travellers in South-East Asia mention these bamboo jackets.

At the end of the last century, bamboo was still being used to make a textile. This was done by cutting the stems across at the joints and boiling them in caustic soda for a lengthy period. Repeated boilings and washings left the fibres exposed and softened, ready to be crushed between rollers. The resulting fibres were carded, combed, and baled. The fibres, which were very soft and able to take dyes readily, were said to be suitable for mixing with wool, cotton, or even silk. Chi Han, writing in the fourth century, referred to a textile made from tender bamboo shoots which were pounded and soaked, then woven into *Chu-shu-pu* cloth. Tougher bamboo was used to make a rather brittle cloth, but both now seem to have disappeared. More recently, bamboo has been used to manufacture rayon.

More common is the use of the bamboo as inspiration for textile design. Colour Plate 18, for example, shows a bamboo leaf design with the branches and stiff leaves made up of short parallel lines of colour. This is the *mee pai bai* or *mee hol* (bamboo leaf) design from north-eastern Thailand. Colour Plate 19 shows another pattern, the *pucuk rebung* (bamboo shoot), from Peninsular Malaysia. This illustration shows the *kepala*, the central or principal panel, of a silk sarong. The main conical figure in the design is a representation of a bamboo shoot, just before the shoot extends. The overlapping leaf sheaves of the bud can be clearly seen. Sometimes very slightly curved edges to the motif suggest fullness and readiness to stretch very rapidly upwards. The design has widespread connotations throughout Malaysia and Indonesia with vitality, human fertility, and good harvests—and is also a symbol of durability. The lozenges of rich reddish-brown between the bamboo shoot motifs are filled with flower shapes.

In Sarawak, an interlocking, comb-like textile pattern is sometimes used which may be based on rattan: as the youngest part of the rattan stem is cleaned of its leaf sheaths, spines with red edges and black tips are laid flat against the creamy white of the young leaf sheath. A pattern which may derive from this is

17. Making a *rawai* 'bodice' (Hedda Morrison).

also found in Dyak bamboo work and pottery.

Another tribal use of rattan in clothing is the *rawai* or *sabit* once commonly worn by women of the Sea Dyak, Kanowit, and other tribes of Sarawak, but now seen only on special festival occasions. The *rawai* is a bodice of hoops of rattan. On to each hoop are threaded sufficient interlocking brass rings to make a cylindrical mail garment which sits on the hips and may cover the bosom. Plate 17 shows a *rawai* being constructed. In Plate 18 each of the young women is wearing a *rawai* as part of her finery.

18. Young women dressed in *rawai* and other finery in Sarawak (Hedda Morrison).

Other Uses of Bamboo and Rattan

Bamboo fibre sandals are made by some Thai hill tribes. The bamboo is soaked in water and pounded to reduce it to fibres, then the fibres are woven to make a thick-soled sandal. As may be expected, these sandals are rough to the feet initially, but become soft and comfortable with use.

Strips of bamboo are covered with paper or silk to make the many fans found in South-East Asia. Patterns may be incised or burned on the bamboo, or designs pierced in it. In a similar way, screens are made from bamboo or rattan and covered with silk or paper or with woven panels of bamboo.

Smoking-pipes are made of bamboo or hardwood. A water-pipe used by people all over South-East Asia consists of a whole joint of bamboo closed at both ends by cross-walls. Plate 19 shows such a pipe being smoked in Sarawak. To make the pipe, two holes are cut, the joint is half-filled with water, and a slender bamboo stem is then inserted into the lower hole, half-way along the joint. Tobacco is placed in this and lighted. When the smoker inhales through the hole cut in the upper cross-wall, smoke from the tobacco is cooled as it passes through the water.

Fine rattans may be woven or plaited together to make belts or straps by which bags are held, such as the bags carried by a strap across the forehead by the Lisu in Thailand. Plate 25, No. 1, shows fine rattan plaiting attached to the Dyak *mandau* or sword. This can also be seen around the gunpowder boxes in Colour Plate 11.

However, the Malacca walking cane is perhaps the most renowned and highest value use of rattan, apart from its use as cricket bat handles and polo sticks. Malacca canes are still traded though are now less fashionable than before. The appearance and stiffness of the Malacca cane are its most important characteristics. What is sought is a stick with no visible nodes (leaf scars). This means the length between scars must be 1 m or so. Only the rattan species *Calamus scipionum* achieves this; in other species 15–75 cm is typical. A cane may

19. Water-pipe for smoking (Hedda Morrison).

be bent to give a crook handle, but to do this without damaging the surface layer around the curve requires great care. Alternatively, a straight cane might be topped with an ivory knob.

Malacca canes have a diameter of 1.5–3.5 cm with a slight taper along the joint. There is a marked 'keel' along part of the length of one side, indicating the base of the 'whip' (see Chapter 7). The surface of Malacca cane is smooth, glossy, and green when first collected but dries to yellow or brown. If bruised during collection, stains are left after drying. The stems may turn black and brittle if not dried promptly, whilst if they are cut too soon they shrivel and crack. It is a local belief that a man will be fortunate if he has an unmarked stick that equals his height (before bending). Sticks made from single joints are known as 'whole-bark' sticks. Cheaper 'half-barks' are made from two canes joined together with a steel pin.

Very pale or black sticks of Malacca cane were considered great treasures in parts of Borneo, equal in value to rhinoceros horn and ivory. Once, when such a cane was found, it had to be presented to the Head of State—the Rajah.

5
Sailing Craft, Fishing, and Hunting

Prahu and Kora Kora

'AT length on April 4th we succeeded in getting away in our little boat of about four tons burthen, in which my numerous boxes were with difficulty packed so as to leave sleeping and cooking room. The craft could not boast an ounce of iron or a foot of rope in any part of its construction, nor a morsel of pitch or paint in its decoration. The planks were fastened together in the usual ingenious way with pegs and rattans. The mast was a bamboo triangle, requiring no shrouds, and carrying a long mat sail; two rudders were hung on the quarters by rattans, the anchor was of wood and a long and thick rattan served as a cable.' Thus Alfred Russel Wallace describes a boat journey in 1860.

There are stories of single culms of bamboo being split in half to make the shell of a small boat. The fourth-century flora of South-East Asia written by Chi Han mentions that one joint of a bamboo known as *Yun-ch'iu*, which grew in Funan (Vietnam and Cambodia), could be made into a boat. *Dendrocalamus giganteus* has the largest known diameter amongst the bamboos, occasionally achieving 25 cm, but even this would mean a tight fit for a slender individual. However, there are nineteenth-century accounts of a bamboo species being used for canoe-building in the Moluccas, with the stem split into halves to make a two-man craft.

Plate 20 shows an interesting craft still used today—a bamboo coracle. Unlike European coracles, where the strength of the craft lies in the wooden frame and not in the canvas or leather cladding, the strength of this North Vietnamese coracle is provided by its woven bamboo skin. Bamboo is traditionally used as a caulking material for coracles and other craft. The bamboo is shredded, then embedded in a putty of lime and

20. Bamboo coracle, North Vietnam (David Goddard).

tung oil. The tung oil evaporates, leaving a solid cement reinforced with bamboo fibres.

Bamboo is commonly used for simple rafts. Being hollow, bamboo stems float high in the water, enabling them to transport downriver the large, heavy logs of hardwood forest trees, such as meranti and teak, which might otherwise sink. The hardwood logs are lashed to the bamboo stems and a bamboo cabin is built aboard. Once the raft reaches its destination, all parts of the raft are sold: the hardwood logs to the sawmill, the bamboo for other purposes. A bamboo raft to transport goods and livestock can be constructed by trained fingers within a few hours. A double layer of bamboo may be used, if necessary. The Vietnamese sailing raft in Plate 21 is a more advanced type of raft that uses the natural curvature of bamboo stems to form a prow.

Another North Vietnamese sailing boat—shown in Colour Plate 20 laden with bamboos—has a truncated triangular sail stretched between two bamboo poles. To close the sail of Madurese boats with triangular sails, the lower boom is lifted towards the other bamboo pole by a rope. The Javanese-style *mayang* rig has a rectangular sail with a bamboo along the top edge supported by a vertical bamboo, and a boom along the lower edge.

Bamboo poles are also used as outriggers for canoes, attached to the narrow, unstable craft to prevent them capsizing. The hollow bamboo pole floats on the surface of the sea more effectively than an outrigger of solid wood. Many other of the accoutrements of South-East Asian boats have been made of bamboo and rattan: ropes, decking, masts, sails, floats, fish-traps, awnings, blinds, beds, cups, buoys on harpoons, anchors, nets, rods, and flagpoles.

Rattan is used in boat-building and also in fishing. Lashing planks together to build boats is an ancient method found throughout Indonesia and further afield. Planks were carved from split trees, every one unique and exactly shaped in three dimensions. Lugs were left projecting and subsequently used as anchorage points for lashing the planks together very firmly.

21. Bamboo raft, North Vietnam (David Goddard).

The Indonesian *orembai* is a narrow passenger boat with upturned prow and stern. Until recently these graceful boats were built with ribs lashed by rattan to internal lugs, a design that dates from the prehistory of South-East Asian boat-building. A very similar method of boat-building was characteristic of early Scandinavian boats and was used in early Viking boats.

Where bamboo or rattan are exposed to water for long periods, they will, of course, eventually rot. This is not a serious matter as they can be replaced easily and cheaply and the labour required is very much less than the effort needed to make a similar article from a more durable material. The war prahu of Sarawak, for example, would be dismantled on return from an expedition and carried into the house as planks. When next required, the boat would be reassembled with new rattan cords. The bamboo coracle of Plate 20 uses rattan to lash the struts which support the upper edge to the bamboo skin. When rattan is stretched taut, it is very stiff, like wire (copper wire was once used in Europe in a similar method of lashed boat-building). The beating of the waves upon the hull may cause the planks to bend and twist, but the strength of the rattan lashings holds the boat together.

Fishing

Cane is used in many ways for fishing. The method used by the Mandar people who occupy many small islands of the Java Sea and the Straits of Makassar is called *rumpun* fishing. The fishing boat used is the *sandeq*, a fast craft equipped with 20–30 long cables made of rattans plaited together, then anchored many miles from land. To attract fish to shaded water, palm leaves are attached to the cable and submerged to a depth of 10 m. Later, the fishing boat returns and a net is lowered around the fish that have collected. Men jump into the water to prevent the fish from escaping the net. The cable is then lifted very carefully out of the water and the fish trapped. They are kept alive in fish tanks in the hull during the return journey to land.

1. Bamboo grove near Zen Buddhist temple at Hokokuji, Kamakura, Japan, similar to bamboo groves found throughout East and South-East Asia (Greg Irvine).

2. Rattan (from W. Veevers-Carter, *Riches of the Rain Forest*, Singapore, Oxford University Press, 1984).

3. Household articles (from P. A. van der Lith, *Nederlansch Oost-Indie*, Doesburgh, 1875).

4. Household articles (from P. A. van der Lith, *Nederlansch Oost-Indie*, Doesburgh, 1875).

5. Architecture at Tangaroeng (from C. Bock, *The Head-Hunters of Borneo*, London, 1881).

6. A Toraja origin-house in Sulawesi, Indonesia, its arched roof made of cleft bamboo strips (from R. Waterson, *The Living House: An Anthropology of Architecture in South-East Asia*, Singapore, Oxford University Press, 1990).

7. Bamboo shoots, cooked and sliced (Brian Pope).

8. Rattan fruits (*Daemonorops micrantha*) (John Dransfield).

9. Thai lunch-boxes (Brian Pope).

10. Cane bird-cages, Thailand (Brian Pope).

11. Gunpowder boxes from Indonesia (Tatiana).

12. Indonesian baskets for carrying pepper (John Dransfield).

13. Samples of popular Thai weaving styles (David Kirby).

14. Seven sages in a bamboo grove, brush pot carved from bamboo stem, Taipei (Lutz collection).

15. Pipe, carved from bamboo rhizome and metal, Thailand, 18–19th century (Lutz collection).

16. Boy with buffalo, carved from bamboo rhizome, China, 19th century, 31.8 cm long (Lutz collection).

17. Bamboo and lacquer brushpot, with rooster and hen, 15 cm high, Japan, Edo period (1615–1868), signed Sankosai (Lutz collection).

18. Bamboo leaf textile design used for silk *mud mee pha sin*, Thailand (from S. Fraser-Lu, *Handwoven Textiles of South-East Asia*, Singapore, Oxford University Press, 1988).

19. Bamboo shoot textile design (*pucuk rebung*) (from G. I. Selvanayagam, *Songket: Malaysia's Woven Treasure*, Singapore, Oxford University Press, 1990).

20. Boat rig (Jeremy Hartley, reproduced with permission of Oxfam).

21. Ceremonial street decoration (*penjor*), Bali (from D. J. Stuart-Fox, *The Art of the Balinese Offering*, Yogyakarta, Penerbitan Yayasan Kanisius, 1974).

22. Children ready to perform folk melodies on *angklung*, Bandung, Indonesia. A *calung renteng* is shown in the foreground (E. Heins).

23. Forest rattan (*manau*), Malaysia (John Dransfield).

24. Spine galleries on rattan (John Dransfield).

Spearing is another method used from the fishing boats. The fish spear rattan (*Calamus curtisii*) is commonly used for this in Malaysia. A spear may be a trident with a rattan handle to which the tines are lashed with rattan. Sea anchors are also made from bamboo mats for steadying boats during fishing trips.

Eggs of flying fish are a valuable catch in the Straits of Makassar and are exported to Japan. To catch the fish and their eggs, a lure is set—a floating bamboo basket with openings surrounded by an inward-pointing fringe of sharp split bamboo spikes. The lure encourages the fish to mate, which they usually do on weed or leaves floating at sea. During courtship the male darts about near the female who is laying her eggs on the basket. In so doing, he may push her into the basket and also get caught himself. The bamboo points prevent them from escaping. Several cane fish-traps are shown in Plate 22. These were in use in Thailand at the end of the nineteenth century, but similar designs are still used all over South-East Asia.

Along the canals and rivers of the mainland, dip nets are mounted to catch a passing breakfast. These also depend on bamboo. The net consists of four poles lashed together at their centres, each one supporting a corner of a large square net about 3–5 m long on each side. The whole net is suspended over the water by another bamboo lashed to a main post and dipped into the water by releasing the supporting central pole. After a period the net is lifted, trapping any fish swimming above it. Smaller, hand-held nets are also used (Plate 23).

It is as fishing-rods that much high-quality bamboo reaches the West. The species *Arundinaria amabilis*, known as Tonkin cane, is particularly noted for its straightness and strength; it can withstand strain for a long period and then return to its original shape. This species is grown specifically for fishing-rods for export from southern China but it seems that the first supplies of this species to the West came from northern Vietnam.

The many kinds of rope found on a boat may be made either from rattan or from bamboo. The simplest type of bamboo

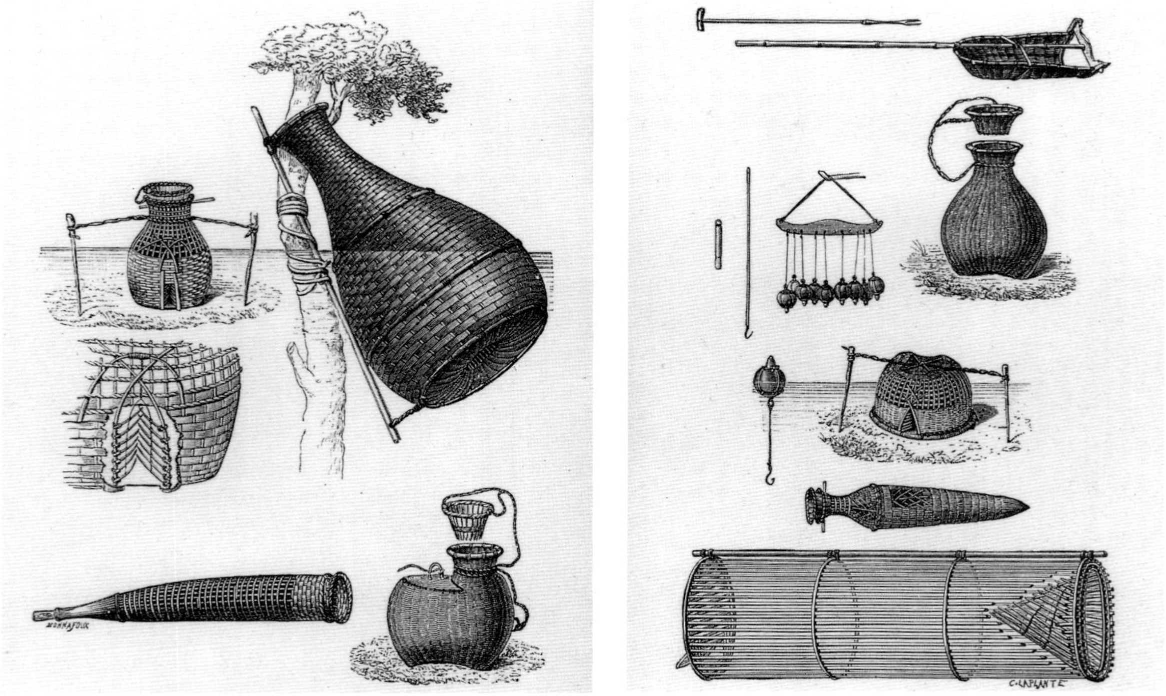

22. Cane fishing equipment (from H. S. Hallet, *A Thousand Miles on an Elephant*, Edinburgh, William Blackwood & Sons, 1890).

23. Hand-held *yaw* or fishing nets, used here in the Central Plains of Thailand (Betty Gosling).

rope is the twisted strand in which strips of bamboo are twisted together into strands with several strands twisted to make progressively thicker ropes. Plaited ropes are even stronger than the twisted kind and more durable, though plaited ropes tend to be finer. Thin strips are plaited together into a braid and braids are then plaited together to give rope. Several ropes can be built up into hawsers of enormous strength, adequate for pulling loaded boats and barges upstream. As the hawser is composed of elements combined together in an open form, water is shed rapidly once the hawser is lifted, and so it dries out easily. Hawsers of this kind are preferred for very hard-wearing uses where the rope will necessarily rub against rock or other hard surfaces. Marco Polo saw such hawsers at work pulling river boats in thirteenth-century China. It is often possible to remove and replace sections of the hawser. Once further repairs are impractical, the hawser is recycled by cutting it into short, thick sections to be used as torches.

Hunting

Bamboo and rattan are both used in a multiplicity of ways for hunting the largest to the smallest animal, and for warfare.

Ling Roth describes how rattan was used in crocodile hunting in the late nineteenth century in Sarawak. River crocodiles were left undisturbed unless they took a human victim, but when this happened, the family or village turned out in force to catch the criminal. Crocodiles eat only rotting flesh, thus when an animal or person is caught alive, a crocodile stores the body until it is suitably high.

To trap the murdering reptile, a wooden hook was firmly attached to the putrefying body of a small animal, such as a monkey, dog, or bird. Attached to this bait-hook was a long length of rattan rope, not anchored but secured to a floating buoy. When the bait was taken, the crocodile carried it to a quiet place to devour his prize in peace. But the buoy revealed the crocodile's location and the reptile would be despatched—after being tied down, ridiculed, and then eviscerated to retrieve any human remains.

At the other end of the scale, rattans are also used in frog-hunting. In the right season, frogs are sought at night by the Dusun of Sarawak by torchlight and are usually found sitting on stones in the middle of a river. They are caught by hand, then threaded on to a rattan cord through their stomachs.

Other animals—deer, pigs, and rats—are caught by snares. A bird-call snare seen by Ling Roth in Sarawak (Plate 24) was made from two lengths of bamboo, one short and thick and another long and slender into which the hunter blew. The instrument made a soft cooing sound, very like the call of the green or puni pigeon. Crouching in a leaf-covered hide with his snare, the hunter blew softly. The sound attracted any pigeon in the area, even enticing it to walk into the hide and be caught.

Larger game may be caught in pitfalls, where a hole is dug and equipped with sharply pointed stakes, then covered with branches and leaves to hide the trap, which may be baited.

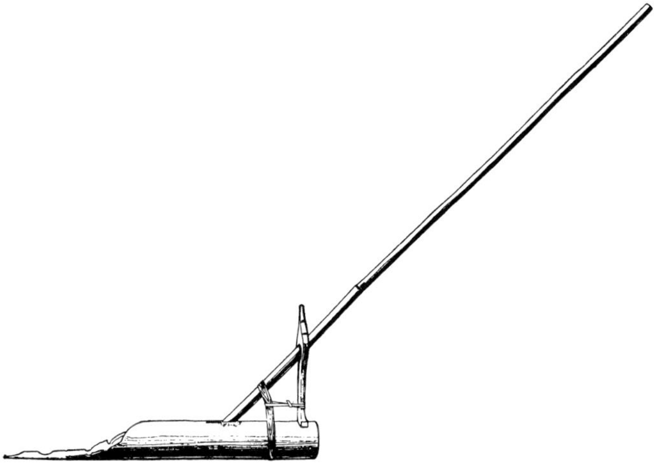

24. Murut bird-call (from H. Ling Roth, *The Natives of Sarawak and British North Borneo*, London, 1896).

There is a stirring account of a man-eating tiger killed by spearing in Sumatra. The assistant of the European naturalist Henry Forbes was killed by a tiger. His remains were buried. The grave was made unusually deep because tigers were expected to return to dig up the remains of the kill. It soon became clear that the tiger was indeed returning to excavate the grave, so a trap was set. After a series of mistakes and misadventures, the snare was set again on the eighth night. A conspicuous gate was left open leading towards the grave whilst other routes were closed by fences. 'Across the opening a thin cord was loosely drawn, connected with a green bamboo some 30 feet long, bent by the strength of several men into a bow at whose extremity a sharp spear was so arranged as to be shot athwart the entrance gate on the release of the bamboo by the tiger pressing with his breast on the twig-like cord in his way.' Next morning it was clear that the trap had been shot and the spear had been broken off. The tiger had escaped with a bamboo spear clear through its body, but was later found dead.

In his account of his Eastern travels, Marco Polo records that in Tibet bamboo was used not to entrap but to opposite effect: to scare away dangerous animals during the night. This was done by tying several green poles of bamboo together and suspending them near to the travellers' camp-fire. As the air inside the internodes heated up, the stems would explode, no doubt frightening not only marauders but the travellers themselves. Bamboo joints are nowadays used in this way on festive occasions, to add to the general uproar and jollity.

Many different weapons have been made from bamboos in South-East Asia. Some are merely defensive. The *ranjau*, for example, is a stake of sharpened bamboo driven into the ground around a rice-field or garden. As only the owner knows the safe path in, others risk nasty wounds to the feet and legs. A similar weapon is the caltrop, once widely used in Malaysian defensive battles.

The blowpipe, or *sumpitan*, was once a common weapon in Malaysia for hunting and is indeed still used by the aboriginal groups scattered throughout the peninsula. It was made of drilled hardwood or bamboo, in particular the *Bambusa wrayii*, which has extremely long internodes, the distance between one cross-wall and the next being up to 2.9 m. From this long and very straight internode can be cut the perfect natural blowpipe. The projectile used may be a small dart cut from rattan and tipped with poison, or a very fine arrow of bamboo, 2.5 mm thick and 20 mm long. Tips of ironwood may be used when large animals are hunted. Where hardwood is used for the blowpipe, as will happen where suitable bamboo species are not available, the central duct is drilled through then it is 'polished' by repeatedly pulling a length of rattan through the hole (Plate 25, No. 9).

The crossbow of the Lahu tribe of northern Thailand, which is used principally to hunt animals, is made partly from bamboo. The stock of the crossbow is made from a hardwood. The bow or crosspiece, however, must bend but be elastic, returning to its previous shape. Bamboo is ideal for this and has great strength to resist the load of the crossbow, which is normally cocked by foot. When it is first made, the crossbow is

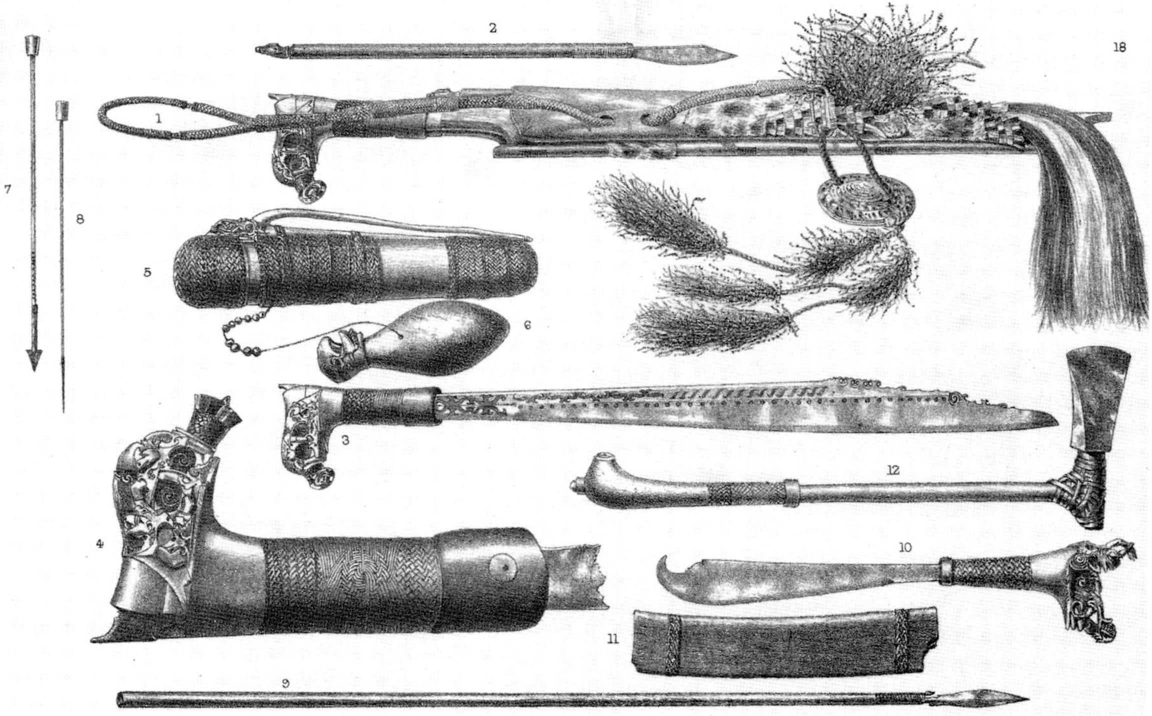

25. Dyak weapons (from C. Bock, *The Head-Hunters of Borneo*, London, 1881).

hung near the cooking fire, and whilst the smoke helps to cure and strengthen the bamboo and the hardwood, any frying fat will contribute to the polish of the weapon. The bolt of the crossbow is cut from bamboo and is not tipped with an arrowhead. The speed and sharpness of the bamboo itself is sufficient to enable the bolt to penetrate even elephant hide. To kill an elephant in this way once earned a hunter great esteem. In cross-section the bolts are oval rather than circular, which means that a greater perimeter of outer skin remains along one side of the bolt, giving strength and helping the bolt to slip across the wood after being struck by the bowstring. For the Lahu, each crossbow is inhabited by a spirit which is propitiated with the blood of every animal slain, to encourage it to provide similar luck on the next hunting foray. The juice of young bamboo shoots are said to provide an antidote to the poison used on the bolts, which may be necessary if a man scratches himself with a bolt whilst out hunting.

Rattan enters the fighting arena in different ways. One rattan, known in Malaysia as *rotan getah*, produces a poisonous resin into which darts are dipped. The naturally barbed whips of rattan have also been used in the punishment of prisoners.

More widely, though, plaited rattan fibres are used to secure weapons. For example, the Dyak quiver (*tolor*) is made from bamboo but traditionally decorated with plaited rattan. In Plate 25, No. 5 shows a *tolor* made from a bamboo joint (30 cm by 5 cm diameter) and decorated with rattan. The quiver is attached to a belt by means of the hook. Hanging from it is the gourd which contains the pieces of pith which are attached to the arrows and which fit exactly inside the blowpipe, making it possible to blow the darts unerringly from the tube. The Dyak *mandau* (literally 'head-hunter') is a single-edged curved blade for fighting and hunting. It is ensheathed in a wooden case made of two sides fitted and held together with plaited rattan (Plate 25, Nos. 3 and 4). The belt that carries the sheath is also made of plaited rattan and from this hang precious ornaments. In his book on Borneo head-hunters, Bock mentioned seeing shells, hornbill beaks, and a human kneecap decorating a belt in the 1870s. The *mandau* itself would have been made locally, in very small-scale production (see Plate 5).

Those who use weapons against each other need armour. No. 10 of Plate 12 shows a rattan helmet used in Borneo (No. 9 is an everyday cap). A protective jacket used in Sarawak was constructed with rattan and bark cloth, then decorated with fish scales. Shields of plaited rattan are also made.

6
Folk Beliefs and Entertainment

THE Chinese philosopher Su Shi summed up the value of bamboo in the eleventh century: 'While man may live without wanting to eat meat, he cannot live without bamboo. Without meat, a man becomes thin; without bamboo he becomes trivial.' In Chinese tradition, the bamboo is likened to a 'slender gentleman'. The Chinese expression for the hollow centre of the bamboo also means 'humble mind', whilst the expression for the rigid knots also denotes 'firm integrity'.

There can be no doubt that the peoples of South-East Asia also identify with bamboo, and to a greater extent than with any other plant. An affinity is perceived between man and bamboo, as best exemplified by the Vietnamese saying, 'The bamboo is my brother', whilst in Sarawak a song likens human life to that of a bamboo leaf. A link between man and bamboo is seen in two Malay proverbs. In one, tactlessness is likened to pulling bamboo 'against its foliage', whilst great friendship is described when two people are said to be 'as inseparable as the bamboo and the river bank'. Further north, in the art of China and Japan, bamboo represents both pliancy and security, since it bends in the storm and shelters the tiger. Sparrows on bamboo stems amid winter snows symbolize friendship.

Bamboos are often associated with purity. Many Burmese folk stories, for instance, begin with the discovery of a girl child in a bamboo culm. Such maidens always grow up to be virtuous and beautiful. In Japan also there are tales of a bamboo node being broken open to reveal a child. A Hindu tale reverses this, with bamboo originating from the girl. A girl of the highest caste, named Murala, was hoodwinked into marriage with a man of the lowest-ranking caste. When she realized that she had been deceived by him, she could not continue living. After praying to Vishnu she climbed a funeral pyre and was cremated. The first bamboo stem grew from her ashes.

In a Malay folk-tale the hero, Khatib Malim Seleman, set off in search of a beautiful princess. He took with him a set of weapons and tools, including a heavy knife, an adze, and a chisel. In the evening he lay down within a bamboo grove and fell asleep. He dreamed that a beautiful princess appeared and cooked food for him, but she disappeared with the dawn.

Khatib knew that she was enclosed within a bamboo joint and tried to set her free, but none of his heavy tools could make a mark on the hard surface of the bamboo. Finally, he tried using a pair of scissors he carried to slice up betel-nuts (many South-East Asian proverbs link betel and love). At last he succeeded in cutting off the top of the bamboo, cleaving the stem in half, and releasing the princess. The two lovers then travelled to Bukit Peraja and disappeared but they can still be invoked to give aid to mortals.

Burmese proverbs reveal human qualities and failings by reference to bamboo. 'Water cannot be forced into a solid bamboo' is said of a dunderhead or one who will not try to understand. When an individual has a well-hidden talent, someone may say that 'the moon is shining in the hollow of the bamboo'. On the other hand, to suggest that someone has a blinkered or narrow viewpoint, it may be said that he sees as through a bamboo stem. Many other sayings use bamboo to emphasize a point. A futile task, for example, may be likened to trying to sharpen a rotten bamboo.

In the Central Visayas of the Philippines, there is a myth which associates bamboo with the origin of man. Sikalak and Sikabay, the first man and woman, were purportedly born from a bamboo cane planted in the first garden by the god Kaptan. They were created in order to tend the garden. They fell in love whilst Kaptan was away on a journey seeking his own estranged wife. Sikabay, the woman, was worried that they should not marry as they were brother and sister, so they consulted the tuna fish, the doves, and the earthquake for their opinion. 'The world must be peopled' the earthquake said and so they married. There are similarities between this tale and that told in the Andaman Islands by the Aka-Bo people who hold that the first man, named Jutpu, was made inside the joint of a large bamboo. After he was born he took clay from the nest of the white ant and fashioned a wife for himself from it.

A bamboo pole is used in several South-East Asian cultures in communications with the spirit world. A field altar is made in Cambodia and in the Central Visayas of the Philippines from the uppermost internode of a bamboo pole which is split into several slats which are then bent outwards to make a cone. The shape is fixed in place by rattan strips woven between the slats. The pole is then set up in a corner of a rice-field. In some cases it holds offerings for the rice spirit at harvest. However, in Cambodian forests this field altar is used for offerings to the *mren-ganval*—small child spirits who guard wild animals and also domestic animals. In the Philippines, Mahagat farmers use a similarly constructed bamboo offering pole in their fields, placing three coconut shells filled with rice in the cone-like basket. In Thailand, the *chaleu*, a bamboo pole topped with a 'three-way lattice' of woven cane, acts as a marker and proclaims ownership of the field. This lattice is encountered in many forms as marker and for protection. In Myanmar (Burma), such a lattice would protect the entrance to a camp as it was believed to 'entangle' evil spirits.

Bamboo traditionally plays a role at many of life's common ceremonies. Following the birth of a child in a Thai hill tribe family, custom prescribed that the afterbirth should be buried in a joint of bamboo. If the child fell sick during its first days, the parents would return to the buried bamboo joint to pray for their child's soul to return to its body.

Bamboo knives are traditionally used for cutting the umbilical cord of the newborn child, whilst in Malaysia, the circumcision knife was almost always made of bamboo rather than metal (bamboo knives are sharp enough to cut elephant hide). At death, bamboo is used to inform passers-by of the family's loss. In Cambodia, a death in the household is considered to expose a house to evil spirits, so an *achar* is consulted and his advice followed closely. When the funeral procession leaves the house, two bamboo poles are erected over it and a series of rites are performed to prevent the spirit of the dead person returning to haunt the family.

Just as the bamboo clump suffers
  at the death of one culm,
So a whole village mourns
  when one person dies.

(opening lines of a Karen death lament)

Bamboo groves typically mark the site of graves, and if the dead person's family is poor, the body may be wrapped in a shroud made from split bamboo. Grave-markers for men are frequently made in the shape of a bamboo shoot. If a man of the Thai hill tribes dies by his own hand, a stake of bamboo will be driven through his heels to stop his ghost from wandering and haunting the living. In a northern Thailand Akha household there will be a shelf holding a bamboo joint closed by a lid of woven bamboo. This is considered to be the home of the ancestral spirits of the family.

But the plant is equally associated with happy occasions. In Bali, bamboo poles mark the great festivals of *galungan* and *kuningan*, when for several days the streets are lined with *penjor*: poles decorated with streamers, arching over the roadway (Colour Plate 21). Roadway decoration by bamboo poles is quite widespread, and 'triumphal arches' are often set up along the route on which an honoured guest or hero will pass. Kurz described bamboo arches being constructed by Malays in the mid-nineteenth century: a principal arch with laterals, all formed of four- to six-cornered columns and joined together with 'skillfully wreathed' trellis work. These customs have survived to be recognizable today. Symbols such as stars and flowers made from thin strips of bamboo and unfurled banana leaves are mounted on the columns which are bedecked with garlands of bamboo flowers intertwined with real flowers.

The use of rockets made from joints of bamboo, packed with gunpowder and decorated appropriately, is another long-standing tradition. These rockets are often made by monks and are used on two occasions in particular: the propitiation of rain gods to encourage them to send rain at the end of the dry season and to mark the death of a monk. Rain-making in Thailand entails several rockets being lit. The height that the principal rocket achieves is said to indicate the amount of rain that will fall: the higher it flies, the greater will be the rainfall. The rockets used at monks' funerals are large and in Myanmar (Burma) are used to set fire to the funeral pyre. Once ignited the rocket trundles across the ground on a small cart, guided by ropes. It hits the pyre and cremation begins. The whistling rocket is another type. A length of bamboo packed with gunpowder is attached to another light bamboo perhaps 5 m long which has been whittled to make a whistle. Several other smaller whistles of various lengths and notes surround the rocket. When it is lighted the rocket climbs rapidly, accompanied by the sound of the whistling bamboos.

Bamboo is a material much used in man's dealings with the spirit world in South-East Asia. It is hard to say whether this is because of the cheapness and availability of the material or because of certain associations with purity and its essential lightness which makes it suitable for use by spirits. For example, Yao tribespeople, like others, use bamboo to construct simple altars or spirit houses. The walls are woven from strips of bamboo and the house frame is made of bamboo culms. Upon a small table or altar are set tiny dishes and cups, carved from bamboo and intended for use by the spirits. 'Silver jewellery' is made from fine strips of bamboo fashioned into chains or other symbols and used to decorate altars. Alternatively, they may be used around a hill tribe village in order either to attract good spirits or appease frightening ones. The Lahu, for example, have a series of symbols made from bamboo, such as woven stars placed on altars. These symbols will be set at the village gate to signal to visitors not to enter when a major ceremony is in progress. Similarly, bamboo symbols warn strangers against entering a house containing a newborn baby, for fear they might cause it harm.

Despite all these signs of affinity between man and bamboo,

a curious superstition exists in Bali which is described by Covarrubias. The Balinese home, or *kuren*, consists of several buildings within a compound, occupied by related families and having a single gate and a single kitchen. A central area is of hardened earth with a few ornamental bushes, usually hibiscus and frangipani. Between the houses and the surrounding wall, fruit trees are grown. Despite its importance to the household, bamboo must not grow within the wall of the compound. Should a clump become rooted within the compound it will not be removed, though indirect attempts will be made to prevent it from thriving. It is the old people of Bali who undertake what is considered to be the dangerous job of removing bamboo and even they will do so only on certain days, for safety's sake. When a bamboo clump is dug out, the first clod of earth is thrown as far as possible. The clod has a symbolic malevolence: should it touch anyone, it is said he will die. Planting bamboo is also considered to be hazardous.

Foretelling the Future

Lightweight and durable, bamboo stems have been used for keeping the seasons of the year. In northern Sumatra, the polished outer surface of a bamboo was engraved with a lunar-based calendar and astrological information so that the *datu*, or diviner, could identify inauspicious days and mark these with the figure of the constellation Scorpio.

Prediction of the future is earnestly sought by many peoples of South-East Asia (and further afield). Bamboo serves for this purpose too, in different ways. A widespread use is fortune sticks. A handful of fine sticks with fortunes and symbols stamped on them are shaken from a container within the temple. Fortunes are then read by the fortune-teller, based on the ambiguous comments incised on the sticks which fall out first.

Bamboo also serves for prediction in Cambodia. Before firing a forest to plant rice, the Brou people will select and split a bamboo lengthwise, then throw the two halves on the ground. The way they fall indicates how successful the firing will be. If both halves fall with rounded outside upwards (or if both inner faces fall upwards), then the firing will not be good. Only if one falls 'heads', as it were, and the other 'tails', then firing will be successful. Not that the recommendation of the bamboo will be followed slavishly, though. If the desired result is not obtained at the first throw, the bamboo will be retrieved and thrown again till the 'right' answer is given.

Toys

Light and cheap, bamboo is the obvious material for children's toys. Bamboo does not splinter (though the outer fibres of a lath may split off) and it lends itself easily to toy-making. Stilts, for example, are easily cut from bamboo for Thai hill tribe children, while hoops may be fashioned from split lengths of bamboo. Strips of bamboo are plaited together to make toy animals such as stylized horses and buffaloes. Amongst the Lahu tribe in northern Thailand, a father will make a tiny crossbow for a newborn son and a tiny spindle for a new daughter. Enlarged versions follow as they grow. A wicker ball is used in ball games across South-East Asia: *takraw* in Thailand, *bola sepak* in Malaysia. This ball is woven using a three-way lattice and has great resilience to kicking.

Kite-flying is a serious pastime for adults as well as children. Thai and Burmese kites are made of rice-paper and stretched by a bamboo frame held in tension by strings and crossbars of bamboo. So successful is this design that kites have been used to carry bombs across enemy lines during battles.

Musical Instruments

Myriad leaves give a thousand sounds
  all are lamentation

(Ou-yang, *c.*1000 BC)

The 'plaintive' or weeping bamboo is designed to be played by the winds of South-East Asia. Indeed, culms of bamboo in a living grove are pierced to allow the wind to blow through. A traveller in South-East Asia in the mid-nineteenth century described the sounds he heard as he approached the villages of Kandingoo: some soft and liquid like the notes of a flute, others deep and full like the tones of an organ. The sound was found to come from a single stem in which all the joints had been perforated with a slit, giving 14–20 notes from a stem. At one time, plaintive bamboos sang at the palace of the Dutch Governor-General in the centre of Bogor's Botanical Gardens.

Several broad groups of bamboo instruments exist, since the shape of the stems and the resonance of the wood lend themselves well to the production of sound. There are percussion instruments that are struck or scraped and very common wind instruments with multiple or single pipes. There are also a few stringed instruments and drums of bamboo. The sound produced depends mainly on the diameter of the bamboo stem used and on the thickness of the woody stem wall, as well as the length of the tube of air that it encloses. The larger the volume of vibrating air, the lower the pitch. The nature of the beaters or strings can also, of course, change the sound.

Percussion instruments used in Java include the *angklung*, a cross between a peal of hand bells and a sliding rattle. The *angklung* consists of three or four separate bamboo joints, each closed at one end by a cross-wall or node, but open at the other end where part of the stem has been removed to leave a tongue of bamboo. The joints are suspended by rattan strings within a bamboo frame and can move from side to side to produce a xylophone-like sound when shaken. Each musician may hold two *angklung*, one in each hand. A complete set of *angklung* consists of eighteen pieces, to give a wide range of notes. Colour Plate 22 shows an Indonesian *angklung* ensemble. The *angklung* is traditionally associated with rice harvesting ceremonies and to arouse soldiers before battle. It was prohibited by the Dutch in Java in order to suppress the threat of rebellion.

Beaters are used with the *calung*, which is made up of 12–18 bamboo joints of selected and increasing sizes attached together

26. *Calung renteng* played at a musical gathering, Sunda, West Java (E. Heins).

with cord to form a 'ladder'. The upper end of the instrument is suspended from a tree or house beam while the lower end is tied to the musician, who strikes the bamboo. The *calung renteng* shown in Plate 26 is built on a frame but uses the same principle. The *gambuang bambu* is similar, though with more joints, to widen the range; the joints are tied by cords to a platform of bamboo.

Bamboo flutes and pan-pipes with 4–7 stops are widespread: some are end-blown, others side-blown. Nose flutes are also known across the region. Whilst some flutes are straight, others like the Timorese *suling*, flare at the lower end. The *suling gambuh* of Bali is an end-blown ring flute about 1 m long. Plate 12, No. 2 shows a Dyak flute of bamboo.

The dewdrops cling
To the green bamboos like pearls.
The wind stirs lazily in the lotuses,
And their pale red petals
Flutter to the earth

As the night creeps on apace,
The twinkling glowworms
Light my path,
And faintly, from a tower
In the east,
Calls a bamboo flute.

Huang Wan Ch'iung, Quing dynasty

Other materials may be combined with the bamboo. Gourd pipes are common and are used, for example, by hill tribe folk in Thailand. Typically, several bamboo pipes are set into holes bored into a gourd or into a hollow hardwood resonating box. The musician blows into the box through a pipe made either from the stem of the gourd or from another bamboo pipe, set apart from the rest. A range of notes is achieved by using bamboo outlet pipes of different lengths, with stops drilled into them. The outlet pipes are often cemented together to give the instrument greater solidity and air-tightness. A similar instrument, described by Bock, is the *kleddi* (see Plate 12, No. 3).

This has six clustered outlet pipes, 25–45 cm long. The longest of these is topped by a short piece of bamboo set at right angles, to modify the tone. Bock describes the pleasing sound as 'organ like', though he complains that when played its music was usually overwhelmed by the vocal contributions of performers and audience.

Mouth harps (or lover's harps) are typically played by courting couples. The *geng-gong* of Bali is one of the many mouth harps in the region. It is a very simple instrument with a vibrating tongue of bamboo held in front of the mouth. By changing the size and shape of his mouth the player can make different harmonics of a low drone. The 'male' harp has a lower note than the 'female' harp, but both types may be played by both men and women irrespectively.

Stringed instruments are more rare, but tube zithers are found in many guises from Madagascar to the Philippines. There is the *calempung* in Java and the *keteng-keteng* in Bali. These are made with two or three strings lifted from the surface of the bamboo joint and stretched over bridges. A hole is pierced in the cross-wall at one end; one string has a bridge at either end. The vibration of a tongue-like strip of bamboo in front of an opening in the joint gives a resonant, gong-like sound when the string is plucked or beaten.

P. and E. Lewis quote the following Karen courtship song, sung by a young man to his beloved:

See the seven-star constellation setting
Untie your loom and come down to the boys
Cut down leaves by the village well
Cut down bamboo by the well of the village
Cut down wood to whittle a harp
Cut bamboo from which a harp can be whittled
Strum it in the house, it is heard in the open
Strum it in the open, it is heard in the house.

The seven-star constellation is the Pleiades, which acts as a seasonal reminder to several South-East Asian peoples, telling them when to clear their land and plant rice. When it sets there is a lull in farming activities, and time for other pursuits.

Drums of all kinds are common in South-East Asia. The *kulkul* is a Balinese split drum made from an internode of bamboo which has a split along its length and is closed at each end by a cross-wall (similar drums are also made from wood). The *kulkul* is slung from a branch and used for signaling and sounding the alarm.

Several of these instruments, like the *angklung*, the tube zither (*keteng-keteng*), and bamboo buzzer, date from the animist, pre-Hindu times in Bali. They are thought to have arisen in the Dong-son culture which flourished on the mainland (Vietnam and Laos) over 2,000 years ago. It is, of course, only metal instruments which have survived from an early date, but wooden and bamboo instruments can be seen amongst the stone reliefs around the temple of Borobudur, built in Central Java in the eighth–ninth centuries AD.

In the making of musical instruments, rattan is frequently used to bind elements together, for example, lacing a skin to a drum. In some cases, however, fine rattans are used as strings on stringed instruments. Writing in the 1880s of the Borneo Dyaks, Bock mentions a stringed instrument called a *djimpai* played like a banjo. This instrument is shown in Plate 12, No. 1. The body is open at the back and the strings are fine split rattan.

7
Natural History of Bamboo and Rattan

WITHIN the variety of environments found in South-East Asia, bamboos and rattans thrive in very different locations. Bamboos prefer open spaces: they grow along river banks, in clearings, and where forest has been cut or burned down. Given the right conditions, bamboos will come to dominate the landscape, covering great tracts of land. Rattan, on the other hand, succeeds best within the rain forest, where it can germinate in the moist surface layer of decaying vegetation and is protected in early life from the full strength of the sun. Rattans require other vegetation for support and shade, and consequently they have developed interesting associations with other rain forest inhabitants.

Bamboos

Although bamboos are vital to the domestic economies of South-East Asia, many questions still remain, perhaps the most intriguing of which concerns flowering. Although it is often believed that bamboos always flower after a long period and then die, the true picture is far more complex. One very common species (*Bambusa vulgaris*) has flowered only sparsely since 1810 and in this time has never set seed. In contrast, some bamboo species flower almost constantly without sign of decline, and between these two extremes other species keep to different timetables for flowering. What 'clock' tells the clump to flower is not known: it could be the slow accumulation of food reserves over the years to a threshold level, or it could be connected with weather conditions. But neither of these explanations solves the riddle of 'gregarious flowering'. When plants of some bamboo species are grown from seeds produced in the same year, they will usually tend to flower together in future years, even in cases where many years have elapsed and where seeds have been taken to very different and distant locations. Seeds produced by one bamboo species in Myanmar (Burma) in 1891 were sent to Calcutta and Dehra Dun (approximately 1 500 km apart), and yet the clumps raised in the two locations flowered for the first time in 1940. In 1970–1, *B. arundinaria* flowered almost throughout India after a gap of forty-five years. In the mid 1950s, a number of thorny Indian bamboos in the Singapore Botanical Gardens all flowered together then died, though no Malaysian bamboo species in the Gardens had been known to flower gregariously.

Bamboo flowers—like all grass flowers—are made up of small, wind-pollinated florets arranged on spikelets. Seeds vary from very small to the size of a small pear, but they appear irregularly and are only briefly fertile so they are less important in spreading the plant than is the underground rhizome. This grows outwards from the clump and sends up new vertical stems (Plate 27). A bamboo stem (or culm) does not stand alone as does a tree in a forest: each stem is merely part of a greater plant which is the clump or grove. The clump may be very long-lived: cutting single stems will not kill the clump and indeed razing or firing them all may encourage new shoots to grow. In some species the new stems shoot at intervals along the rhizome. This group are called 'runners' (see Plate 27) and give spacious groves, such as that shown in Colour Plate 1. Another group—more common in South-East Asia—develop into a dense and impenetrable thicket, as shown in Plate 1. These 'clumping' bamboos are used to form a protective hedge or living wall around a house or village, made all the more impassable by spines growing from the leaf sheaths which surround the stem. Plate 28 shows how 'clumping' bamboos grow.

A bamboo culm does not grow stouter from year to year like a tree though it may live for several decades. The culms produced by a young rhizome will be relatively slender. As the years pass and the rhizome builds up food reserves, each year's new culms become thicker and stronger.

27. Runner bamboo (from A. and C. Rivière, *Les Bambous, végétation, culture, multiplication*, Paris, Société d'Acclimatation, 1878).

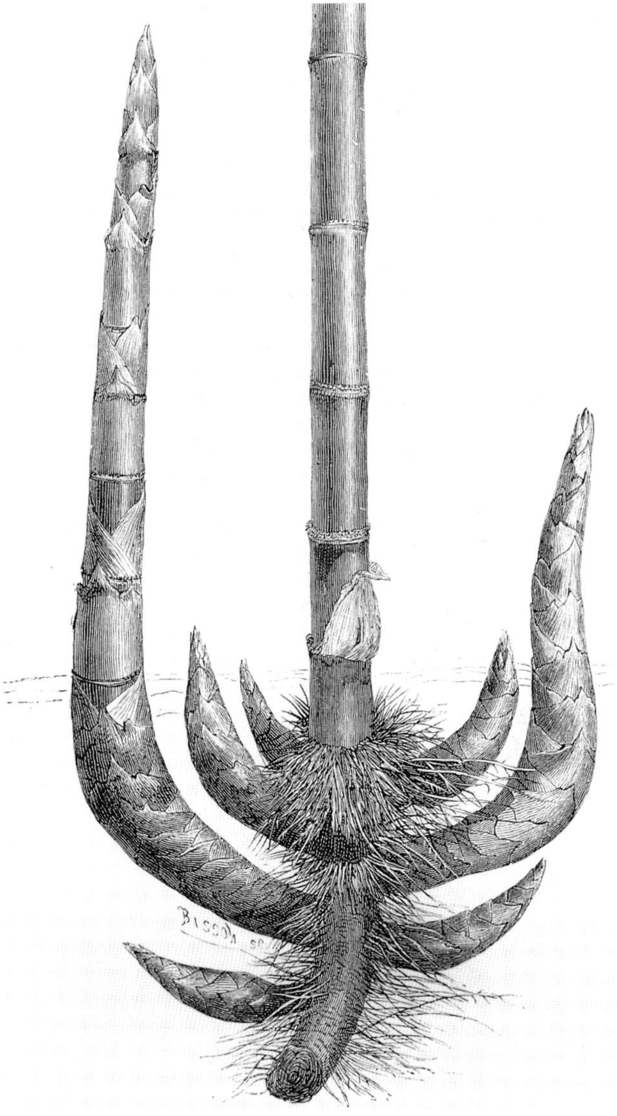

28. Bamboo clump (from A. and C. Rivière, *Les Bambous, végétation, culture, multiplication*, Paris, Société d'Acclimatation, 1878).

A new bamboo shoot is thick and conical, packed with energy and nutrients to enable it to extend rapidly. This rich food source should be very vulnerable to grazing animals, but it is protected by prickles on the leaf sheaths and sometimes by toxins. The new shoot waits for suitable weather conditions, then extends upwards as an unbranched cane. In monsoonal areas the new canes spurt skywards in the rainy season, though growth is steadier where rainfall is distributed evenly throughout the year. When the cane is at about half its final height, it grows at its fastest rate. It is said that at this stage bamboo can be seen growing, reaching 3 cm per hour, though this peak is short-lived. After a few days, the growth rate gradually relents.

When the stem has reached a half to two-thirds of its final height, the stem starts to branch from a small knot of buds on one side of the stem (see Plate 11). The branches grow in the same way as the main stem, by extending a series of joints.

Leaves finally appear and for the first time the stem can provide its own food supply from sunlight and contribute to the survival of the mother clump. Bamboo foliage is not dense. The leaves are long and oblong or lance-shaped, and are held out stiffly from either side of the branch.

A young bamboo–
how tall it has grown
without the slightest help in the world!

Bamboo cultivation may involve merely harvesting whatever becomes available; alternatively, fertilizers may be added and the rhizomes thinned or even moved. Where bamboo shoots are harvested for food, the farmer will throw fresh earth over the rhizomes every year to encourage shooting and to protect them until they are ready for harvest. Sunlight spoils the flavour of fresh young bamboo shoots.

Bamboo is normally propagated by planting a length of bamboo with some attached rhizome. Vegetative reproduction is not always easy but occasionally success is unexpected. Across Central America, for example, bamboo canes have taken root when they have been used to support bananas.

This final note on bamboo cultivation is taken from Kurz, writing in 1876, and quoting a Captain Sleeman: 'In the rains of 1835, my bamboos at Jubbalpore had not thrown out their shoots at what I considered the usual time and I asked my gardener the cause. He replied: "We have had no *thunder* yet; as soon as the thunder comes, you will get shoots." I asked him what possible connection there could be between the claps of thunder and the shooting of the bamboos. "God only knows", said he, "but we know that till the thunder comes, the bamboos never shoot well". The thunder came and certainly the gardener's theory seemed to me to be confirmed by a very steady and abundant shooting of the bamboos.' Kurz suggested that the connection might lie with the addition of nitrogen to the soil by lightning, although the probable size of this contribution does not seem enough to fully explain the phenomenon.

Rattans

Wallace believed that the rattans 'greatly improved the appearance of the forest as seen from the coast, for they vary the otherwise monotonous tree tops with feathery crowns of leaves rising clear above them, and each terminated by an erect spike like a lightning conductor'.

The most common species of rattan in South-East Asia are those of the *Calamus* genus which has seventy species in Malaysia alone. Other major groups are *Daemonorops* and *Korthalsia*. The life-cycle of rattan is quite different from that of bamboo. Rattans do not spread by underground rhizomes as do bamboos but only by seed, so flowering and seeding must be regular if the plants are to survive. A thick carpet of rattan seedlings is often to be seen near an established rattan stem, but clearly there are many losses on the way to maturity.

Rattan fruits are something of a forest delicacy, favoured by birds, monkeys, and many other animals. The fruits are picked and the seeds spat out or passed through the gut and voided. Small piles of seedlings may be found growing together. At first only a rosette of leaves is visible above ground, enclosing the growth point or palm heart. The palm heart is surrounded by protective sheaths throughout its life, but it is attractive and tasty to animals. Its loss often means the death of the plant.

The 'lightning conductor'—the new sword leaf—protrudes from the tip of the stem, tightly furled (Colour Plate 23). This leaf opens out and slowly bends backwards. As it does, it may touch another plant or tree and then become closely attached to it by the hooks along the leaf's mid-rib. The mid-rib is extended as a 'whip' armed with more fearful hooks which enable the rattan to reach the canopy by clinging to the growing trees. The leaf sheaths which protect the palm stem are also armed with hooks. Some rattans grow as single stems or groups of only three or four, but others may have fifty stems. When a supporting tree dies, or where a tuft fails to find a support, coils of rattan may subside to the forest floor.

In the rain forest home of the rattans, close associations exist between plants and insects. Plants trade food and shelter for services like pollination and seed dispersal by animals. Plant and animal species have evolved together and mutual dependencies have developed. Rattans are frequently inhabited by scale insects. These small, flat, greyish-brown insects suck sap from the stem, but they do not come to the rattans of their own accord. They are transported there—we could almost say pastured there—by ants. Ants are unable to collect sap from the rattans as their mouths are not the right shape, but they can 'milk' the scale insects for a sugary liquid. The scale insects are introduced by the ants into natural spaces within the rattans such as the ocreas of the canes—a sort of natural 'legging' around the stem. Ants also live in galleries made by parallel rows of spines encircling the stem (Colour Plate 24) and protect their living quarters with papier mâché made partly from the tissues of the rattan itself.

Both the scale insects and the ants thus derive food and shelter from the rattan—but how does the rattan benefit? Benefit it must: the ocreas and spine galleries seem to have evolved specifically to accommodate ants and scale insects.

This is an account by the naturalist E. Corner of a day in the

Malaysian rain forest: 'We pause in the excitement of a first expedition through the sweating glades beneath the dipterocarps and a faint rustle fades into the trees. It comes again and vanishes as widely. We think of snakes, wasps, bees and look around, but can discern no obvious source of the mystery. Yet now it is repeated, right, left and on all sides as a series of tiny rattles through the bushes into the canopy. An uncanny feeling shivers our clammy bodies that we are watched by a host of invisible onlookers. But is this snake underfoot any more than the trailing stem of an old rattan? We bend down in relief to examine it and, just as the sound commences, a small ant at a hole in an ocrea beats its mandibles on the dry cover. Instantly a rattle comes from the box, to be taken up by the next and the next along the stem into the trees. We are standing on an old clump of *Korthalsia* [rattan] and all its stems are rattling. The sentry has alerted the soldiers and the alarm is transmitted from post to post as a sort of "action stations" to repel boarders up to the crown in the tree tops. We lift the stem and out rush the ants with ferocious bites.'

Elephants, pigs, and squirrels are very partial to rattan heart. It may be that where spines and thorns have failed to protect the plant, the ants provide a final line of defence. Rattans with resident ants are less frequently destroyed by marauding animals than ant-free rattans.

Harvesting rattans

Rattans are not well loved by botanists in the field. Says Corner, 'Unattached whips, so fine and tenacious, are an ever present hazard in the forest of South-East Asia . . . to both man and beast. One needs to be a pachyderm for immunity. To disengage the whips it is necessary to back and in backing, other whips may be encountered; the unwary may become incredibly and most painfully entangled.'

Most rattan canes are collected from the rain forest by men who often do this arduous and poorly paid work to supplement their main livelihood, often a small rice farm. Indonesia pro-

29. Kelabit rattan collector at an informal gathering in Sarawak, miming the task of gathering rattan (Hedda Morrison).

duces 90 per cent of the world's traded rattan, Thailand and Malaysia almost all the rest.

The rattan collector sets off into the forest armed with hand tools. He must first locate a suitable rattan plant in the forest to which he may return periodically as it puts out new stems over the years. Then he must pull down as much as possible from the canopy, or collect the coils which may lie on the ground where a large tree has fallen. In Plate 29, a Kelabit collector is seen demonstrating his job of pulling down rattan in a comical mime to entertain friends at a party in Borneo.

Apart from ants' nests, dislodging a stem may also bring

30. 'Runti rotan': removing the outer surface from rattan by pulling the cane through a hole (John Dransfield).

down wasps' nests, dead limbs of trees, clumps of epiphytes, and, of course, the waving barbed whips of the rattan itself. The oldest part of the stem may be too well anchored to be pulled down whilst the youngest parts will be too soft to be used. Between these extremes is the saleable rattan. First, the leaves and leaf sheaths must be removed. This is done in the forest by pulling the rattan over or round a tree trunk to rub them off, perhaps with the aid of a parang. If slender enough, the rattan may then be coiled up and carried back to the village. Large rattans are heavy with dilute sap and are left to drain before carrying. Plate 30 shows the process of '*runti rotan*', preparing the cane by removing the outer surface of the rattan by pulling it through a hole.

Though harvesting rattans from the natural forest is the origin of much commercial rattan, there is also widespread planting of rattan by villagers for their own needs, and increasingly rattan is being grown as a cash crop, especially in Malaysia, often where forest has been felled and is growing back, or in old rubber plantations. In the future, much of the world's rattan will probably come from rattan estates and small-holdings on which rattan is cultivated by farmers.

Bibliography

Austin, R., Ueda, D., and Levy, D., *Bamboo*, New York, Weatherill, 1970.

Bock, C., *The Head-Hunters of Borneo*, London, Sampson Low, 1881; reprinted Singapore, Oxford University Press, 1985.

Burkill, I. H., *A Dictionary of the Economic Products of the Malay Peninsula*, Singapore, Governments of Malaysia and Singapore, 1966.

Chi Han, *Nan-fang ts'ao-mu chuang, A Fourth Century Flora of South-East Asia*, Introduction, translation, and commentaries by Li Hui-Lin, Hong Kong, Chinese University Press, 1979.

Corner, E., *Natural History of Palms*, London, Weidenfeld & Nicholson, 1966.

Covarrubias, M., *Island of Bali*, Alfred A. Knopf, Inc., 1937; reprinted Kuala Lumpur, Oxford University Press, 1972, and Singapore, Oxford University Press, 1987.

de la Loubère, S., *A New Historical Relation of the Kingdom of Siam*, London, 1693; reprinted Kuala Lumpur, Oxford University Press, 1969, and Singapore, Oxford University Press, 1986.

Dransfield, J., 'A Manual of the Rattans of the Malay Peninsula, Malaysia', Forestry Department, *Malayan Forest Records*, No. 29, 1979.

Forbes, H. O., *A Naturalist's Wanderings in the Eastern Archipelago*, New York, Harper & Brothers, 1885; reprinted Singapore, Oxford University Press, 1989.

Jumsai, S., *Naga: Cultural Origins in Siam and the West Pacific*, Singapore, Oxford University Press, 1988.

Kurz, S., 'Bamboo and Its Use', *Indian Forester*, Vol. 1, No. 3, pp. 219–69, 1876.

Lewis, P. and Lewis, E., *Peoples of the Golden Triangle*, London, Thames and Hudson, 1984.

Ling Roth, H., *The Natives of Sarawak and British North Borneo*, 2 vols., London, Truslove & Hanson, 1896; reprinted Kuala Lumpur, University of Malaya Press, 1968.

McClure, F. A., 'Bamboo in the Economy of Oriental Peoples', *Economic Botany*, Vol. 10, pp. 335–61, 1956.

———, *The Bamboos: A Fresh Perspective*, Cambridge, Massachusetts, Harvard University Press, 1966.

Newman, T. R., *Contemporary Southeast Asian Arts and Crafts*, New York, Crown Publishers, 1977.

Nieuhof, J., 'Mr John Nieuhoff's Remarkable Voyages and Travels into ye best Provinces in ye West and East Indies', published in the second volume of Awnsham Churchill's *A Collection of Voyages and Travels . . .*, London, 1704. Originally published in Dutch, Amsterdam, 1682. The second part of the 1732 edition of this work was reprinted as *Voyages and Travels to the East Indies 1653–1670*, Singapore, Oxford University Press, 1988.

O'Flaherty, W. D., *Other People's Myths*, London, MacMillan, 1988.

Otsuka, R. Y., *Selections from the Lutz Bamboo Collection*, Denver Art Museum, 1979.

Polo, M., *The Travels of Marco Polo*, translated by T. Waugh from the Italian of M. Bellonci, London, Sidgwick and Jackson, 1984.

Purseglove, J. W., *Tropical Crops: Monocotyledons*, 2 vols., London, Longman, 1968.

Robinson, G. W. (trans.), *Poems of Wang Wei*, Harmondsworth, Penguin Books Ltd., 1973.

Waley, A., *One Hundred and Seventy Chinese Poems*, New York, Alfred A. Knopf Inc., 1929.

Wallace, A. R., *The Malay Archipelago*, London, Macmillan & Company, 1869; reprinted Singapore, Oxford University Press, 1986.

Winstedt, Sir R., *Malay Proverbs*, London, John Murray, 1950.

Index

References in brackets refer to Plate numbers; those in brackets and italics to Colour Plate numbers.

CHINA
SZECHUAN
PROVINCE
MYANMAR
(BURMA)
Vientiane
Rangoon
THAILAND
LAOS
VIETNAM
Bangkok
CAMBODIA
Phnom
Penh
ANDAMAN
ISLANDS
STRAITS OF MALACCA
BRUNEI
M A L A Y S I
Kuala Lumpur
SARAWAK
Malacca
SINGAPORE
BORNEO
SUMATRA
I N D O
JAVA SEA
Jakarta
MADURA
Bogor
JAVA